Professional Ethics in Construction and Surveying

This textbook responds to the increasing demand for practical, industry-aligned, ethical practices in quantity surveying, construction management and related AEC professions. *Professional Ethics in Construction and Surveying* addresses how existing ethical standards can pragmatically be applied to both private and contracting practice, with case studies aligned with the ethical requirements of the main professional bodies. After an introduction to ethics, the authors present real-world situations where the minimum legal and contractual requirements necessitate the combination of professional judgement and ethical decision making. They outline how such situations arise, then address how decisions can and should be made that are in keeping with the moral, contractual and CSR requirements, with cases covering the building lifecycle from procurement to handover. Consequently, the book brings together ethical theory, existing worldwide ethical standards and the requirements of the RICS, the CIOB and the ICES, with the authors' experiences of examining candidates for entry into the professional bodies.

The result is a professionally focused textbook aimed at vocational learners (at both undergraduate and postgraduate taught levels) and practitioners in construction, engineering, architecture and the wider built environment.

Greg Watts, BSc, PGCert, PGCAP, EngD, MRICS, FHEA is the Director of Quantity Surveying at the University of Salford, UK. He is also an RICS chartered quantity surveyor with 15 years' experience in the construction industry and has published numerous conference and journal papers.

Jason Challender, PhD, MSc, FRICS, FAPM, FAHE is the Director of Estates and Facilities at the University of Salford, UK, and a member of its Senior Leadership Team, responsible for overseeing a large department of approximately 350 estates and construction-related staff. He is also a construction academic researcher with two books and ten academic journal and conference papers published, all of which have been dedicated to his studies on collaborative procurement in the construction industry. He is a Fellow of the Royal Institution of Chartered Surveyors (RICS), the Advance Higher Education (AHE) and the Association of Project Managers (APM). Furthermore, he is a Board Director of the Royal Institution of Chartered Surveyors and the North West Construction Hub.

Anthony Higham, BSc, PGCE, MSc, PhD, MRICS, MCIOB, FHEA is a chartered quantity surveyor and chartered construction manager with over 20 years' experience spanning industry and academia. Anthony is currently Head of Construction and Management at the University of Salford, UK, and an Associate Tutor at the University College of Estate Management. Anthony has been professional review assessor for the Chartered Institute of Building since 2012 and has previously chaired the Science and Technology ethics committee at the University of Salford, UK.

Peter McDermott is Professor of Construction Management in the School of the Built Environment, at the University of Salford. He is a founder member and Joint Coordinator of the World CIB Working Commission (W92) into Construction Procurement. Peter has published widely in peer-reviewed journals on the subjects of procurement and supply chains and social value and industry development.

Professional Ethics in Construction and Surveying

Greg Watts, Jason Challender,
Anthony Higham and
Peter McDermott

Routledge
Taylor & Francis Group
LONDON AND NEW YORK

First published 2021
by Routledge
2 Park Square, Milton Park, Abingdon, Oxon OX14 4RN

and by Routledge
52 Vanderbilt Avenue, New York, NY 10017

Routledge is an imprint of the Taylor & Francis Group, an informa business

British Library Cataloguing-in-Publication Data
A catalogue record for this book is available from the British Library

Library of Congress Cataloging-in-Publication Data
Names: Watts, Greg (Gregory N.), author. | Challender, Jason, author. | Higham, Anthony, author. | McDermott, Peter, author.
Title: Professional ethics in construction and surveying / Greg Watts, Jason Challender, Anthony Higham and Peter McDermott.
Description: Abingdon, Oxon ; New York, NY : Routledge, 2021. | Includes bibliographical references and index.
Identifiers: LCCN 2020049906 (print) | LCCN 2020049907 (ebook) | ISBN 9780367354169 (hardback) | ISBN 9780367354190 (paperback) | ISBN 9780429331855 (ebook)
Subjects: LCSH: Construction workers--Professional ethics. | Building--Professional ethics. | Quantity surveyors--Professional ethics. | Quantity surveying--Professional ethics. | Construction industry--Moral and ethical aspects--Great Britain.
Classification: LCC TH159 .W375 2021 (print) | LCC TH159 (ebook) | DDC 174/.9624--dc23
LC record available at https://lccn.loc.gov/2020049906
LC ebook record available at https://lccn.loc.gov/2020049907

ISBN: 978-0-367-35416-9 (hbk)
ISBN: 978-0-367-35419-0 (pbk)
ISBN: 978-0-429-33185-5 (ebk)

Typeset in Bembo
by Taylor & Francis Books

Contents

Illustrations

Figures

Tables

Boxes

Foreword

As the sector which designs, constructs and maintains the built environment, the need for ethical practice is of paramount importance in the construction industry. For all construction and surveying professionals, being aware of, and maintaining ethical standards are, however, far more than simply ensuring that legal compliance is achieved. It is about doing what is correct and right, being honest and accountable, and holding yourself and those around you to account for delivering higher standards and expectations. Having high ethical standards could be described as the distinction of a true construction industry professional. Ethical behaviour is a prerequisite for professional body membership, and a key underpinning element of continuing professional development.

Despite the plethora of guidance on how construction professionals can be ethical practitioners, there is sometimes a gap between the ethical standards promoted and the ethical behaviours undertaken. Whether it is because of production or commercial pressures, or the complex, one-off and temporary nature of many projects, ethical practice can take a back seat to other concerns. The inherent temporality of construction can also create ethical issues for those involved in unfamiliar situations and working with unfamiliar collaborators. However, given that construction activity operates almost wholly in the public eye, it is inevitably scrutinised much more than many other industries, presenting real challenges for those responsible for evidencing consistent, ethical practices across complex supply networks.

With their vast experience of the construction industry, the authors are well placed to address these challenges with this textbook that provides an explanation of the pragmatic and practical application of ethical standards. The book traverses the full range of ethical considerations for the construction professional, from describing the nature and significance of ethics, the role ethical practice plays at every stage of the RIBA Plan of Work, and the varying approaches to organisational ethical practice, to how social value and corporate social responsibility behaviours can demonstrate the effectiveness of ethical practices and policies. This textbook offers a comprehensive introduction to, and application of, ethical practice for construction and surveying professionals.

Andy Dainty
Dean, Architecture, Building and Civil Engineering
Loughborough University, UK

Preface

Due to its size, the construction industry is a major contributor to the UK economy, with its impact extending into all aspects of society. The actions and behaviours of its professional workforce, therefore, can have both positive and negative ramifications, depending upon their ethical behaviour. What is considered and accepted as ethical, however, may sometimes not be consistent among all construction professionals. We all have our own interpretation and perception of ethics, and of which behaviours are ethically acceptable, and which are not. While ethics can mean different things to different people, there is a vast consensus in the construction industry, and wider society, regarding ethically acceptable actions and behaviours. Within the construction industry, ethics can be considered from a personal perspective, for each and every professional, or from a wider business perspective, with the ethical practices of organisations. Ethical behaviour can also be governed from both a legal standpoint, through legislation and case law, and from a professional body standpoint, through the publication of best practice guides, ethical standards, and general rules and regulations. There exists, however, a gap in how these ethical frameworks and guidance documents can be applied and evinced in a pragmatic and meaningful manner in the daily operations of construction professionals.

The aim of this book is to provide an ethical overview of the construction industry so that all professionals who operate within it, and those planning on entering the industry in the future, are aware of, and can apply, ethical standards to their own behaviour and the wider practices of their organisations. When enacted correctly, ethical consideration can ensure fair and robust procurement is undertaken, the potential for disputes is minimised, careers can progress, discrimination can be removed, skills can be developed, bad practices can be reduced, and the industry and its professionals can continue to build upon their existing ethical practices and standards.

This book was written in response to two main issues experienced by professionals in the construction industry. First, this book serves as a guide as to how and why ethically correct decisions should be made in the variety of situations that professionals face. Second, professionals of all hierarchal levels operating within the industry experience difficulties when attempting to communicate ethics, especially when trying to align theoretical frameworks with pragmatic examples. Hopefully, this book illustrates how ethical considerations can create a proactive framework to inform behaviours and ensure professionals are aware of the correct ethical decisions to make in a variety of situations.

1 Introduction

1.1 Context to the discussion on professional ethics

The UK Government's Construction Strategy 2025 identifies the strategic significance of the UK construction industry to the wider UK economy. The industry currently consists of over 280,000 businesses, providing employment for over 3 million people (HM Government, 2013). In addition to the economic benefits, the industry also extends and improves the fabric and facilities of society, thus meeting the collective and individual needs of an array of internal and external stakeholders (Calvert, 1995; Cohen and Grace, 1998). Notwithstanding this premise, over the last decade the morality of the industry and the direction of its moral compass have come under increased scrutiny. In 2009, tender practices within the industry, associated with bid rigging or collusive tendering, came under scrutiny from the Competition and Markets Authority, resulting in fines in total of £129.2 million against 103 contracting organisations. The 2017 Taylor Review of modern working practices raised serious concerns about payment practices in the UK construction industry, especially the morality of supply chain management practices which have been seen to impose payment terms of up to 120 days on sub-contracting organisations. In 2018, the collapse of Carillon led society to question the ethical practices of the construction industry. Furthermore, in 2018, Dame Judith Hackitt's independent review of Building Regulations and Fire Safety, following the Grenfell Tower disaster, raised major questions about contractors' use of 'value engineering' to reduce costs at the expense of safety and once again led to society questioning the morality of the construction industry. There have been calls for change from within the industry. Farmer (2016) adopted medical metaphors, describing the construction industry as a sick patient in urgent need of treatment if it was to survive.

Despite the many ethical and commercial challenges facing the industry, construction firms obstinately continue to legitimise their actions through the claim of inclusion of corporate social responsibility within their reporting structures, often using this policy apparatus as both a rebuttal of the negative press and a panacea response to clients' increasing demands for socially responsible project delivery. While internally, those tasked with delivering promised social deliverables remain unconvinced and thus not committed to the ethical and moral promises made.

For construction, design and other professionals operating within the construction industry, the incalculable value of human life demands nothing less than the highest moral considerations (Mason, 1998). In philosophy, ethical behaviour is seen as ascribing to oneself good or right behaviour. Bommer et al. (1987) advocate a long-standing tradition that exists in ethical philosophy as prescribing 'ethical behaviour' as behaviour which is shown to be objectively and morally correct, via an appeal to a theory of morally correct action, and state that it is 'ethical' precisely because it is the behaviour which is required by ethical theory. Yet the high-profile negative practices exhibited by some construction companies are arguably the result of the unethical management practices of staff. However, professional bodies, such as the Royal Institution of Chartered Surveyors (RICS), provide construction professionals with a means of gaining integrity and respectability through their affiliation. Yet ethical statements produced by professional bodies tend to be theoretically informed, produced in isolation from one another and developed with little consideration for practical application. Accordingly, when aspiring members attempt the peer review process associated with chartership, they struggle to translate the theoretical pronouncements into real-world practice examples. The authors' experiences of examining candidates for one of these institutions over a number of years have revealed that while candidates found demonstrating technical and management competence relatively straightforward, the same candidates would regularly struggle to demonstrate their commitment to ethics, morality and codes of conduct in a robust and appropriate way.

The approach taken by this book will respond to the increasing demand for practical and industry-aligned, ethical practice in quantity surveying and construction management. More specifically, the book will address how existing ethical standards can be pragmatically applied to both private and contracting practice, with case studies and example scenarios aligned with the ethical requirements of the main professional bodies for the quantity surveying and construction management professions. Furthermore, the book will provide coverage of those real-world situations where the minimum legal and contractual requirements continue to require quantity surveyors and construction managers to demonstrate professional judgement and ethical decision-making. The textbook will outline how such ethically problematic situations arise, often resulting from differences in expected behaviours between international companies, from conflicts between legal and moral dicta, or from the mismatch between the business goals of the client and the contracting organisation's and society's perceptions of them as social facilitators. The textbook then addresses how decisions can and should be made in such circumstances that are in keeping with the spirit, which is often beyond the minimum, of the legal, contractual and corporate social responsibility (CSR) requirements. Consequently, the book will bring together ethical theory, existing worldwide ethical standards and the requirements of the Royal Institution of Chartered Surveyors (RICS), the Chartered Institute of Building (CIOB) and the Chartered Institution of

Civil Engineering Surveyors (ICES). Practical advice will also be provided on how the largely theoretical ethical principles and guidelines adopted can be applied to all quantity surveying and construction management disciplines at both national and international levels.

Construction contracts are clear on the responsibilities and duties of each party, and what will occur if one party fails to complete a particular contractual duty. However, such contractual clauses are often used in isolation of the larger project requirements, therefore, leading to unfair practices. One example includes the NEC4 clause 50.5, whereby 25 per cent of a payment can be withheld if a programme is not included with the contractor submission. While this clearly applies to all parties, on occasions it may not be possible to include a programme, and not doing so will have no impact on any other construction activity. Yet 25 per cent of monies will be withheld, which has a huge detrimental effect on the contractor and possibly causes cash flow issues that would then negatively impact all stakeholders. Another example is of quantity surveyors not paying retention when due, or intentionally hiding contractor requirements within a large contract, in the hope they are missed by the contractor, yet they are deemed to be contractually included, and then demanded at a later date, to the detriment of subcontractors' cash positions. Situations of potential bribery and ethical dilemma with no immediately clear right or wrong answer regularly occur within the construction industry and are increasingly scrutinised by professional bodies in their assessment interviews.

This book shifts the focus from traditional textbooks on ethics towards practice and those professionals working towards professional qualification, in the form of chartered membership of one of the established professional bodies. The reason for this is that previous publications have failed to address the practical application of ethics in the day-to-day realities of professional practice. For instance, there has been little focus on the specific requirements of the professional bodies and how readers looking to use a text to guide preparation for the ethics aspects of competency-based assessment would translate theoretical ethical constructs into actual real-world practice.

The book is also intended for experienced practitioners who wish to use the text to meet their continued professional development (CPD) requirements. Furthermore, it will provide a much more professionally focused textbook aimed at vocational learners (at both undergraduate and postgraduate taught levels). Using the RICS and CIOB ethical standards (among others), the book seeks to pragmatically apply such standards to real-world scenarios. This is designed to offer readers practical guidance in real-world practice, addressing the key limitations of previous books on professional ethics which have predominantly focused on academic theory. Chapters will be supported by extensive case studies; detailed tasks and supporting seminar questions. Some of the chapters will provide formative assessment tasks, allowing readers to assess their own progress through the text. All tasks and questions will be supported with suggested solutions. The textbook provides an ideal preparation tool for those seeking chartership of construction and surveying professional bodies where ethics forms a core and fundamental area of assessment.

1.2 The structure of the book

The book covers a diverse range of subject areas with the overarching focus on professional ethics. This includes areas such as socially responsible procurement, the increasing importance of corporate social responsibility (CSR), recent changes to European procurement directives and changes to the ethical requirements and guidance of professional bodies. The book begins with the history, nature and importance of ethics. Chapter 2 is specifically focused on the overriding question of 'why is ethics so important to society?'. It will attempt to define what ethics are but, in this pursuit, will articulate the ongoing dilemma of the different definitions of ethics. Such differentiation could pose a problem in the quest to understand and interpret what constitutes ethics and ethical practices. The book will also delve into the area of why ethics are important in Chapter 3 and will discuss the aims and benefits for individuals, organisations and society at large of ethical principles and values. In this regard, examples of unethical practices in the construction industry will be specifically discussed with the consequential effects and fall-out from these in terms of reputational damage and loss of public trust. The important issue of how to recognise unethical practices and dilemmas over what is right or wrong will be explored and how to address such adversities will be discussed. This is designed with the intention of changing cultures, addressing existing bad practices and improving working practices, especially in the construction industry, for the better.

Chapter 4 then considers ethical considerations specific to the UK construction industry and discusses the importance and significance of the construction industry to global economies, including the UK, while highlighting the complex nature of the sector. Complexities around the bespoke nature of building projects will be covered, alongside the temporary organisational aspects of the project procurement in Chapter 5. Furthermore, risks associated with the dynamic and fragmented environment that encapsulates the construction industry will be identified. This chapter will also include the roles and responsibilities of different members of project teams, including clients, project managers, architects, quantity surveyors, engineers, main contractors and subcontractors. It will consider these different roles from the perspective of ethical considerations and how they vary, depending on the individuals. As a consequence of these differing perspectives, dilemmas around ethical decision making will be discussed and ethical investigation models will be identified, which seek to address such dilemmas in assisting construction practitioners to decide what is right or wrong. In addition, the notion of building life cycles will be forensically analysed and the competing dilemmas of capital costs versus operational costs.

Chapter 6 discusses the role of professional bodies in the construction industry and how these professional bodies impact and influence the relationships between client, consultants and contractors. How professional body standards relate to ethical principles and practices is discussed as well as the sanctions professional members potentially face for any breaches of their respective professional bodies code of conduct.

In Chapter 7, the application of ethical frameworks in practical construction industry-based scenarios as they relate to both the pre-design and design phase of construction projects will be examined. As ethics at this stage of projects often relates to financial decision making, it will cover financial project performance from the consultant's perspective, and present the benefits of the early integration of main contractors into project teams. Ethical concerns and dilemmas will be established, and real-world project case studies will be explored. These areas will hopefully demonstrate the complexity and challenge of determining the boundaries of legal and ethical practice and the challenge of addressing unethical practice from the general failings in project governance.

Chapter 8 will articulate why there needs to be ethical principles for those professionals working in the construction industry and will explain the importance of such principles. It will also highlight the challenges for ethical teaching, given that there are no universal standards around the world, or between different professional institutions, on what constitutes good ethical practices. It will explain codes of conduct in regulating professional ethics, what these comprise and how they are applied to improve practices and behaviours. In this regard, it will discuss the benefits of having codes of conduct and how they vary between different professional bodies and institutions. Penalties and sanctions for non-adherence and breaches of professional codes of conduct will be covered, alongside the implications for public trust and confidence of violations of ethical regulations. Alongside this, the importance of maintaining public trust will be discussed, given that the construction industry has traditionally been regarded as poorly performing in the past with low levels of client satisfaction. Comparisons will be made of many professional bodies in their approaches to professional ethics, standards and behaviours and the differences between their codes of conduct examined.

Chapter 9 will examine the ethical dilemmas construction professionals are likely to encounter as projects are concluded and the completed building is handed over to the client. This is an area seldom covered by other construction management textbooks and covers the period where the building enters its longest phase 'in use' or of 'occupancy'. In this regard, the book will examine some ethical dilemmas associated with life cycle costing, repurposing, extension, refurbishment or indeed demolition when the building life cycle comes to an end. It will seek to explain the role and involvement of other professionals, including building surveying and facilities management colleagues, as the building responds to use and begins to deteriorate, thus necessitating maintenance interventions. This, of course, does not remove the ethical dilemmas; it will simply introduce a series of new ethical considerations for examination.

To conclude, Chapter 10 offers some final points for consideration. It will summarise the main findings of the book and address where future research is required to tackle some of the inadequacies of current practice. Furthermore, it will reflect on some of the challenges that are faced by the construction industry and offer pragmatic solutions to achieving best practice.

References

Bommer, M., Gratto, C., Gravander, J. and Tuttle, M. (1987). A behavioural model of ethical and unethical decision making. *Journal of Business Ethics*, 6: 265–280.

Calvert, S. (1995). Managing stakeholders. In J. R. Turner (ed.), *The Commercial Project Manager*. London: McGraw-Hill.

Cohen, S. and Grace, D. (1998). *Business Ethics: Australian Problems and Cases*. Melbourne: Oxford University Press, Australia.

Farmer, M. (2016). *Modernise or Die: The Farmer Review of the UK Construction Labour Market*. London: Construction Leadership Council.

HM Government (2013). *Construction 2025. Industry Strategy: Government and Industry in Partnership*. London: HM Government, pp. 23–25, 61–71.

Mason, R.R. (1998). Ethics: A professional concern. *Civil Engineering*, 68(12): 63–66.

2 The history, nature and importance of ethics

2.1 Introduction

This chapter is specifically focused on the overriding question: 'Why is ethics so important to society?'. As context, it will attempt to define what ethics are but, in this pursuit, will articulate the ongoing dilemma of the different definitions of ethics. Such differentiation could pose a problem in the quest to understand and interpret what constitute ethics and ethical practices. Furthermore, as a foundation for the discussion on ethics, the origins and history of ethics will be explored, ranging from the philosophies and teachings of Socrates in the fifth century BC to modern-day doctrines on ethical frameworks. In this regard, historical theories will be explained and contextualised as the basis for ethical teaching over the centuries.

The chapter will also delve into the area of why ethics are important and will discuss the goals and benefits for individuals, organisations and society at large of ethical principles and values. In this regard, examples of unethical practices in the construction industry specifically will be discussed and the consequential effects and fall-out from these in terms of reputational damage and loss of public trust. Finally, the important issue of how to recognise unethical practices and dilemmas about what is right and wrong and how to address such adversities will be explored. This is designed with the intention of changing cultures, addressing existing bad practices and improving working practices, especially in the construction industry.

2.2 What are professional ethics?

To address the issues of professional ethics, particularly applied to the construction industry, first of all, one needs to understand what ethics are and what constitutes ethical or non-ethical practices. Professionals are bound by a set of attitudes, principles and character dispositions that govern the way their profession is practised, and this is commonly referred to as 'professional ethics'. Such ethics are not confined to clients but, according to obligations, are also owed to colleagues and members of the public at large.

Theories of ethics emerge from a philosopher's perspective and can be categorised as:

- metaethics, relating to where ethical values and principles emerge;
- normative ethics, relating to moral standards of conduct;
- applied ethics, involving examining controversial issues (Internet Encyclopaedia of Philosophy, 2010).

Ethics has been described in general usage as:

- 'the philosophy of human conduct with an emphasis on moral questions of right and wrong' (Helgadottir, 2008);
- 'the system of moral values by which the rights and wrongs of behaviour are judged' (Rosenthal and Rosnow, 1991);
- 'a moral philosophy that involves systematizing, defending and recommending concepts of right and wrong behaviour' (Wikipedia, 2019).

Alternatively, ethics could also be defined as:

> the systematic attempt to make sense of individual, group, organizational, professional, social, market and global moral experience in such a way to determine the desirable, prioritized ends that are worth pursuing, the right rules and obligations that ought to govern human conduct, the virtuous intentions and character traits that deserve development in life and to act accordingly.
>
> (Petrick and Quinn, 2008, cited in Helgadottir 2008)

A general definition of ethics was provided by Carey and Doherty (1968, cited in Poon 2003) concerning the philosophy of human conduct with an emphasis on moral questions of right and wrong. Notwithstanding this, professional ethics has been described specifically concerning the expectations of responsibility, competence and willingness to offer quality services to the general public. There is sometimes confusion about what constitutes business and professional ethics. Accordingly, it would be helpful to examine the difference between the two. To offer clarity in this regard, business ethics are primarily focused on conduct. They revolve around ethical questions related to whether certain actions can be deemed good or bad, right or wrong, virtuous or vicious, worthy of praise or blame, reward or punishment. It seems that defining professional ethics is not an easy and straightforward task. To reinforce this position, Uff (2003) stated that 'in any event, it would be difficult today to pin down ethics to a particular definition in the current climate of change in matters of professional accountability and transparency'. Notwithstanding this challenge, professional ethics can be defined as 'a set of moral principles or values' (Trevino and Nelson, 2004), whereas others have described them as a system of norms to deal with both the morality and behaviour of professionals in their daily practice (Abdul Rahman et al., 2007). Alternatively, Fewings (2009) described professional ethics as the application of values to society and a range of different perspectives, whereas Bayles (1989, cited in Abdul Rahman et al.,

2010) advocated that professional ethics revolve around a systems of norms wherein day-to-day behaviours and morality can be managed and regulated. Notwithstanding these alternative descriptions, when considering professional ethics, it is important to understand what 'profession' means. Profession in this sense could be defined as 'the possession and autonomous control of a body of specialist knowledge, which, when combined with honorific status, confers power upon its holders' (Uff, 2003). This is quite an arduous definition and as such it is probably better to consider professionalism from the perspective of providing a 'service' to clients. In this regard, a profession can be linked to a group of individuals who can provide specialist services, including knowledge, to their clients. Professional ethics therefore could be defined as a system of standards and norms to enable the behaviour and morality of those practising a profession to deal with their duties and job responsibilities. However, according to Abdul Rahman et al. (2010), professional ethics does not always apply solely to individuals but can also be extended to organisations and institutions. Furthermore, one could argue that professional ethics are not solely for the benefit of clients but should be extended to the public. Moreover, they encompass qualities and duties associated with responsibility, and competence to abide by established standards, rules and behaviour.

When considering professional ethics, Lere and Gaumnitz (2003) described them as an approach between professionals and experts, with the clients as lay people. Conversely, when we examine professional ethics specifically within the construction industry, the Royal Institution of Chartered Surveyors (RICS) defined them as 'a set of moral principles extending beyond a formal code of conduct' (RICS, 2010). In terms of how such moral principles should be applied in the workplace, the code of ethics for project managers stresses that 'it is vital that Project Managers conduct their work in an ethical manner' (Walker, 2009).

According to Rogers (1911), ethics revolves around the well-being of human beings. Well-being in this sense signifies the permanent realisation of goodness enacted by individuals and takes into account:

- the nature of individual good;
- the nature of social good;
- the relationship between individual and social good;
- the freedom of the will;
- the ethical worth of positive morality;
- the relationship between good and pleasure;
- the nature of virtue (in antique ethics). In this sense, virtue is linked with a person's character and applied to their motives or actions. In this context, it is generally accepted that a morally virtuous person is one who abides and respects the moral codes laid down by honour, mercy, industry, temperance, and charity. Conversely, the opposite of virtue is associated with vice.

- duty and moral obligation (in modern ethics);
- the ethical motives that exist for people to pursue social good or whatever is morally right.

Another challenge when describing and defining ethics is that they have two distinct levels, namely, micro and macro, relating to different aspects of the organisational environment. According to Chang (2005), micro ethics relates to personal issues and relationships with individuals and deals with issues such as personal integrity, honesty, trust and transparency. Alternatively, macro ethics deals with the wider aspects of actions affecting and impacting on society, the environment and the reputations of organisations (ibid.). Macro ethics can be related to ethical issues that have a significant and high-level impact on the reputation of the industry or the public at large and are predicated on the notion of doing good things for the wiser society.

Clearly, the aforementioned alternative descriptions and definitions highlight that there are many differences in explaining what ethics are, which has given rise to problems of ambiguity and meaning. This view was supported by Vitell and Festervard (1987), who advocated that complications arise, and ethical dilemmas exist as there is no universally accepted definition of ethics. Perhaps this is one of the problems when considering professional ethics in the construction industry on a global scale, which could be giving rise to different interpretations of ethics for construction professionals and what they mean in practice in different countries. This is arguably of particular importance when considering the construction industry, as the understanding and views of many different professional bodies, association and organisations and their members do not always match. The RICS Professional Ethics Working Party (RICS, 2000) accepted that this potential ambiguity and inconsistency could create problems for the profession. Perhaps, therefore, a common framework for managing ethical dilemmas and thereby improving ethical standards is required, which could address this problem. This view is supported by Liu et al. (2004), who argue that professional ethics is based on the subjective nature of principles, standards and values, which vary between different sectors of the industry, and thus a more consistent approach is accordingly required.

2.3 A brief historical and theoretical perspective on ethics

The known history of pure ethics or moral (ethical) theories began with some of the ancient Greek philosophers, such as Socrates, Plato and Aristotle. After recovery by the English Positivists, ethics were the main area for discussion during medieval times in Europe. In more modern times, throughout the nineteenth century, ethical ideas and theories were debated robustly throughout Europe. Evolution concepts emerged in the physical sciences as well as the development of ethics during this time and were supported by the likes of Darwin, Comte and Spencer. Some of the ethical precepts of the main early philosophers are summarised below.

2.3.1 Socrates (fifth century BC)

Socrates was the founder of the Science of Ethics and professed the following moral beliefs and virtues:

- Learn your passions within your own soul and control them to reach wisdom.
- He who knows must act accordingly.
- No one voluntarily follows evil.
- Virtue is knowledge.
- Vice can only be because of ignorance.
- Only by self-knowledge can freedom be acquired.

Socrates believed that religious figures such as Zeus did not act in a moral way at all times. Accordingly, he advocated that people are not sure if behaviours are good because it pleases the gods or if it pleases the gods because it is good. Furthermore, he professed that some difficulties, such as geometry, can be solved through data, whereas others, including the justice system, are moral issues.

2.3.2 Plato (fifth–fourth century BC)

Plato defined the social good and their relationships in his famous book, *The Republic*. He advocated the four virtues belonging to the state, which were wisdom, courage, moderation and justice. In this regard, justice was classed as the highest virtue and included the other three virtues. He claimed the highest good was in the form of ideas and reason in the universe and defined the notion of a mortal body and an immortal soul. He advocated that good people are the ones in whom knowledge, emotion and desire are in perfect harmony with their souls. He believed that people should give others the benefit of the doubt and not assume they have done wrong without sufficient evidence. Plato exclaimed that 'no one knowingly harms himself or does evil things to others because that would harm his soul'. Furthermore, Platonic ideals were predicated on justice being regarded as intuitive and arising from individual perceptions of forms.

2.3.3 Aristotle (fourth century BC)

Aristotle defined political science as the highest of all sciences, as everything else was aimed at the good of the state. Furthermore, 'social good' was considered by him above 'individual good' and this was related to individual actions attaining goodness for society. He defined well-being as the activity of the soul and stated it was in accordance with virtuous practices. Aristotle advocated that people should balance emotion and rationality and that ethics should be predicated on moral choices and reason guiding actions, with objectivism at its heart. Furthermore, his philosophies revolved around the notion that everything in the known universe has a beginning, end function, skill and purpose.

2.3.4 Early Christianity

Early Christianity focused on having a sense of one own personal goodness and morality and professed that this was the foundation of ethics. This stemmed from the underlying premise of 'love thy neighbour as thyself and do unto others as you would have them do unto you'. The early Christians also believed that quality is the basic principle of what they referred to as the 'golden rule'. Conversely, they proclaimed that anarchy was due to people not wanting to be ruled as this implied to them inequality in society.

In addition to the above, Table 2.1 presents other historical perspectives and schools of thought regarding ethics.

Table 2.1 Other historical perspectives and schools of thought around ethics

Ethical perspective	*Description*
Hindu ethics	Hindu ethics were largely based on the qualities of virtue, truthfulness, self-restraint and inner purity. They were also predicated on the understanding that ethics cannot always be derived from first principles.
Daoist ethics	These were predicated on passivity leading to self-realisation and that human nature is basically good
Thomas Aquinas (thirteenth century)	Thomas Aquinas practised the notion that the reasoning of individuals naturally propels them to act in virtuous ways. However, he stressed that not all acts of virtuous acts necessarily flow from natural inclination.
Thomas Hobbes (seventeenth century)	On the basis of social contract theory, Hobbes professed that not all things are objectively evil or good and people base good on what pleases them
David Hume (eighteenth century)	Hume argued that ethics are not always derived from rationality but more so on the way people feel. Furthermore, he professed that morality is inherent within the majority of the population. This, he argued, allows them to determine what is evil or good but conceded that rules are required, due to the limitations of individuals.
Jeremy Bentham (nineteenth century)	Bentham studied and practised Unitarianism and he preached that those actions which provide goodness for people could be considered to be righteous acts. He was a believer that 'the ends justify the means' in most cases and that the morality of a particular act could be judged not by its intentions but by the consequences. As such, ethical practices could be evaluated on the acts of people and specifically their behaviour which achieves goodness for others. He concurred that altruism in society should be as effective as possible to avoid or reduce the suffering of others and achieved by the most effective way available to them.

2.4 Historical theories as frameworks for ethics

In addition to the above historical perspectives, the following theories have presented themselves in the past as theories as frameworks for ethical teaching.

2.4.1 Deontology

Deontology was commonly perceived as a framework predicated on the notion that human beings have a propensity and duty to follow rules on ethics. Accordingly, there was a common understanding that people should treat others as they would expect to be treated themselves. Deontology also revolved around the belief that good intentions and good will are still good, even if their results do not achieve good results or consequences.

2.4.2 Social contract theory

Social contract theory was first introduced by Thomas Hobbes in the sixteenth century. The theory was grounded on the premise that people in their natural environment and habitat are constantly fearful of violence and death. In this regard, human lives were viewed as 'solitary, poor, nasty, brutish and short'. In an attempt to escape the miseries of these aforementioned predicaments, people under social contract theory doctrines were encouraged to relinquish powers they had over others in exchange for the protection afforded to them from a strong authoritarian central medium. Protection in this regard was viewed as a social contract. Altruism and cooperation were regarded by social contract theory as vital to create a system of ethics based on collectivistic and social ideologies. The individual handed over power to the state and thus society was formed.

2.4.3 Virtue ethics

Virtue ethics were predicated on the basis that people should behave in accordance with principles of virtue and be compassionate to others. Virtue is regarded in this sense as bringing pleasure and a peace of mind through doing the right thing and therein attaining the highest value.

2.4.4 Cognitivism

Cognitivism stemmed from the notion that propositions that are expressed from ethical statements can be either true or false. Conversely, non-cognitivism theories professed that ethical statements were the articulation and expression of opinions and emotions that were not always true to life. Accordingly, these expressions were considered to be more subjective than objective and neither true nor false in nature.

2.4.5 Scientific ethics

In the 1950s, Robert Merton advocated that there was scepticism regarding scientific theories and therein they were always open to challenge and questioning. In this regard, he philosophised that transparency and disclosure of results and data were not always practised, with scientists trading intellectual property rights for esteem and recognition. He advocated a standard of Universalism to bring about truth, predicated on devising pre-established criteria. Following a similar belief, Bruno Latour in the late twentieth century was equally as sceptical of science, society and technology. He opined that 'scientific facts' were made up of social constructs designed to overwhelm the public by marshalling enough supporters and users. Over many years previous theories about the universe have been proven to be wrong and those previously dismissed as wrong have proven to be right. Latour also suggested that knowledge of science was sometimes flawed as it is an artificial product of economic, political and social interactions in a competitive environment.

2.5 Concept and purpose of ethics

According to Guttmann (2006), ethics can be designed to provide the mental powers to individuals to enable them to overcome fleeting passing instincts and therein enable them to choose good preferences over bad. Furthermore, Guttmann stated that they should facilitate one's thought process in order that decisions emerge from the conscience of individuals to 'do the right thing'. Being ethical in this regard, he argued, should involve adopting moral codes, linked to values and principles. According to Guttmann's doctrines, these are designed to ensure that rules and practices are consistently applied across a broad range of business situations. Such ethical practices should be conducted to benefit not only clients but in society at large in all day-to-day business affairs. Guttmann argued that it is important, especially for new construction professionals, to fully recognise what is required of them. This is predicated on the basis that professional conduct is not something that can be ignored and the way they conduct themselves is a major element in their recognition. Furthermore, Guttmann professed that individuals should be taught to know right from wrong in their decision making and be confident and assertive enough to uphold their actions and stand up to unethical behaviour.

According to Vee and Skitmore (2003), one of the important aspects of ethics is 'personal ethics'. This can be described as treating other people with the same amount of honesty and consideration as they themselves would expect to be treated. Some professionals would agree that these personal ethics are more important for their clients than other individuals, including other members of the public.

2.6 Goals of professional ethics

According to Fewings (2009), the goals of professional ethics can broadly be classified into two main categories: inward-facing and outward-facing goals, and these are listed below.

2.6.1 Inward-facing goals

- Providing support and guidance for people to behave and act in an ethical manner, especially in instances where they are faced with dilemmas and pressures.
- Creating common rules for people and organisations and outlining their responsibilities and duties to act and behave ethically.
- Deterring individuals from acting unprofessionally and behaving in an unethical way by creating an environment in which people are encouraged to report unethical behaviour and where sanctions are identified.
- Reducing internal conflicts.
- Creating standards of acceptable behaviour between colleagues, employees, employers, associates, clients and the general public.

2.6.2 Outward-facing goals

- Protecting vulnerable populations who could be harmed by the activities of a profession.
- Responding to previous cases of unethical conduct by professions.
- Creating institutions that are resilient in the face of external pressures.
- Serving as a platform for adjudicating disputes between professions and between members and non-members.
- Providing a platform for the evaluation of professions and as a basis for public expectations.
- Establishing professions as distinct moral communities and worthy of autonomy from external control and regulation.
- Creating a basis for the development of trust and enhancing and protecting the good reputation of professions.

2.7 The importance of professional ethics for the construction industry

In order to discuss professional ethics in context, it is important to understand why it is important for the construction industry. Philosophers have debated for many centuries on the importance of ethics in society, which formed the basis of how the study of ethics awareness, education and standards has developed in the modern world. In this regard, perspectives have changed over the centuries, but ethical principles are largely built upon values, behaviours and trust in the society in which we live and work. Ethics has become a 'hot topic' in many sectors in recent years and the construction industry is no exception to this agenda. Notwithstanding this premise, reported cases of immoral and unethical practices and in some cases fraudulent criminal activities have given rise to some high-profile cases which have tainted the construction industry and put it in a bad light. This has had an overarching detrimental effect on the reputation of the industry and created a distinct lack of public trust in the construction profession and organisations. There have been reports of sleaze and corruption and ethical improprieties linked to collusion and regulatory breaches. Ethical teaching in applying professional standards and behaviour in the construction industry is vitally important to turn the tide and respond and address such unethical practices.

In the past, there have been views that ethics are a vital and essential practice requirement as they engender the general public's trust and preserve their employers' interests (Abdul Rahman et al., 2007). Construction professionals who act in an ethical manner will enhance their performance which will increase the success of their projects. In addition, for construction professionals to survive, public confidence is necessary, which depends on ethical conduct and professional knowledge (RICS, 2010). Accordingly, this could suggest that any compromises on ethics could jeopardise the service delivery and damage public perception and the image of the construction industry.

One of the other problems the construction industry has faced for many years lies in its reputation and the public's general view of the industry (Robson, 2000). This has largely arisen from negative press coverage, particularly in the UK, on unethical practices and behaviour uncovered in the past. From data published by the Building Research and Information Service, it would seem that contractors, compared to other construction professionals, have a worse reputation for unethical behaviour. This assertion is predicated on a higher number of legal disputes emanating from construction industry behaviour and practices over successive years (Vee and Skitmore, 2003). Research studies by Liu et al. (2004) suggested that developers and contractors especially place little emphasis on ethical codes compared to other construction professionals, such as architects and surveyors. This may therefore suggest that contractor-led organisation, such as the Chartered Institute of Building (CIOB), should develop and promote cultural change to improve standards and raise awareness of ethical procedures. Irrespective of whether the contractors are at fault, reputational damage to construction professionals may ensue when controversy over projects, as a result of their contractors' wrongdoing, emerges in the press. There have been many examples where clients have suffered because of the practices carried out by their appointed contractors. The example of Carillion's demise in 2018 in the UK led to speculation that public sector organisations, who awarded framework contracts for schools and hospitals, had been negligent in not undertaking more robust financial checks in their due diligence procedures.

A possibly more controversial explanation for unethical practices in contracting could be the relatively high turnover of construction company personnel and the perceived motivation of greed for increased profits (ibid.). Construction companies are frequently accused of being concerned only with short-term economics or simply blinkered to the interests and well-being of their clients and associates in the industry. After all, why would construction companies become unduly concerned for their clients and practices, when in most cases they move from one job to another in short succession? Arguably, this is a misunderstood area in construction management. It can produce adversarial management styles geared to aggression and deception rather than a professional approach underpinned by integrity, honesty, transparency, fairness and trust (Walker, 2009). Accordingly, this may explain why the importance and practice of professional ethics do not often feature very highly on the agenda of some building contractors. Construction professionals should be aware of such negative aspects and manage their selection processes to appoint contractors who have demonstrated social and moral responsibility in the past.

A previous example of controversy was related to construction workers being added to a database and blacklisted by up to 40 major national construction companies in 2003. It was reported that this could have excluded many individuals and companies from employment without their knowledge. This followed previous failures of standards and ethics related to the Office of Fair Trading's investigation and report of September 2007 which found that 103 construction firms had colluded with competitors in bid-rigging to secure construction contracts. This bid-rigging practice is further discussed by Vee and Skitmore (2003), who explained that tendering has traditionally been the prime area of concern for the construction industry brought about by unethical practices on the part of both contractor and awarding client.

The aforementioned cases involving main contractors' misdemeanours does not infer by any means that other construction professionals, including clients, always comply with ethical codes. There have been cases in the past which involve construction clients using their powers as 'paymasters' on projects, to pressurise their professional consultants and main contractors to accept fixed price commissions and tenders which do not reflect reasonable margins for them to survive. Such win-lose scenarios, devoid of a partnering/teamwork philosophy, can lead to reputational consequences for construction professionals under such scenarios. Clearly this is damaging to the construction industry at large in terms of public confidence and trust in the sector. While speculation exists as to the existing law on such matters, these cases certainly highlight the debate on whether ethical standards and codes of conduct within the construction industry are being maintained.

The Chartered Management Institute (CMI) state that ethical standards apply equally to the personal behaviour of individuals, as they do to organisations if their actions impact on society at large (CMI, 2010). Furthermore, Trevino and Nelson (2004) described the importance of ethical standards from a personal perspective, stating that staff need to feel that they are acting with integrity and in accordance with best practice. They explained that individuals and professionals prefer to work for ethical organisations to feel good and be proud of the work they undertake. On a managerial level, organisations are responsible for the ethical or unethical practices of their employees under the law. For this reason, it is important for employers to set out standards for acceptable and non-acceptable behaviour and practices for their staff (ibid.). This is especially relevant for managerial staff who should not abuse the authority and power of their positions and must conduct their roles legally and legitimately. Walker (2009) argued that power can be a positive or negative force in its illegitimate form. Power in its positive form can be used to further the objectives of the organisation, whereas in its negative form, it may be used to achieve personal objectives, which do not match organisational objectives.

One of the benefits of ethics for organisations revolves around trust, especially where firms are heavily reliant on their reputation to conduct business and gain new work (Trevino and Nelson, 2004). Weiss (2003) outlined that business ethics deals with three basic areas of managerial decision making. The first concerns choices made regarding the law and whether to follow it, the second relates to choices on economics and social issues outside of the law and the third is the priority of self over the company's interests. Despite these assertions, Cowton and Crisp (1998) argued that economies depend on the profit motive, and the pursuit of profit need not be perceived as immoral. To offer assistance to practitioners concerning what is moral or immoral, Cowton and Crisp stated that one needs to consider: (1) respect for core human values; (2) respect for local traditions; and (3) the belief that context matters, to decide what is right and what is wrong.

Arguably, ethics can create the bedrock on which all of our relationships are built; it is about how we relate to our surroundings and it is not about the connection we have but the quality of that connection (Trevino and Nelson, 2004). Another important aspect, which is at the heart of ethical and moral issues for construction professionals, is the ethical dilemmas and decision making. It is recommended that there is a need for sensitivity to ethical dilemmas to avoid actions being taken without being aware of potential ethical issues.

2.8 Ethical principles and codes for construction professionals

Owing to increasing concerns in many high-profile cases, including those previously referred to in this chapter, demonstrating dishonesty and corruption, it is important for construction professionals to commit to and encourage project teams to comply with sustainable ethical principles. Codes of ethics have provided an indicator that organisations and institutions take ethical principles seriously as they outline expectations for all personnel with regard to ethical behaviour and intolerance of unethical practices (CMI, 2013).

Relationships between construction professionals and the professional consultants and contractors they appoint rely on professional ethics and trust, especially since fee agreements cannot accurately specify all the financial contract contingencies for possible additional services (Walker, 2009). The main reason why the public rely on members of professional bodies relates to them giving advice and practising in an ethical manner (RICS, 2010). Accordingly, the RICS has developed eight ethical principles to assist their members in maintaining professionalism:

1 Honesty
2 Openness

3 Transparency
4 Accountability
5 Objectivity
6 Setting a good example
7 Acting within one's own limitations
8 Having the courage to make a stand.

In order to maintain the integrity of the profession. members are expected to fully commit to follow these values. Furthermore, a 'Code of Ethics Checklist' published by the CMI sets out that ethics is particularly relevant to maintaining the reputation of an organisation and inspiring public confidence in it (CMI, 2013). For this reason, codes of ethics should reflect the practices and cultures which construction professionals want to encourage for their respective organisations and project teams. This is supported by the CMI who advocate that: 'A code of ethics is a statement of core values of an organisation and of the principles which guide the conduct and behaviour of the organisation and its employees in all their business activities.'

Arguably the main deficiencies of codes of ethics have arisen from the notion that there are no universal standards and accordingly they vary between countries and different sectors in the building industry. Boundaries and barriers created by fragmentation and differentiation within the construction sector have possibly deterred any common frameworks of professional ethics from emerging in the past (Walker, 2009). This is an area that demands more attention through multinational dialogue across all areas of the construction sector. One attempt to address unethical behaviour in this way comes from the Global Infrastructure Anti-Corruption Centre, which has published a guide with examples of corruption in the infrastructure sector to assist practitioners. It sets out potentially criminal acts of fraud, which include collusion, deception, bribes, cartels, extortion or similar offences at the stages of pre-qualification and tender, at project execution and in dispute resolution (Stansbury, 2008). Furthermore Abdul Rahman et al. (2007), from studies conducted on construction professionals, published rankings for the top eleven most frequent unethical practices and these are presented in Table 2.2, and these include under-bidding, bribery and collusion.

Codes of conduct and ethics will be discussed further in Chapter 6.

2.9 How do construction professionals recognise unethical practices?

The next issue and potential problem relates to what constitutes unethical behaviour in practice. As previously highlighted, there is no universal theory of ethics, with different cultures existing within the construction industry and this creates problems and dilemmas in knowing what is ethical or non-ethical (Liu et al., 2004). Clearly this reinforces the need for construction professionals to have a consistent approach to professional ethics which can be applied across

Table 2.2 Top eleven most frequent unethical practices

Ranking of most frequent acts of unethical conduct	*Ranking of frequency*
Under-bidding, bid shopping, bid cutting	1
Bribery, corruption	2
Negligence	3
Front loading, claims game	4
Payment game	5
Unfair and dishonest conduct, fraud	6
Collusion	7
Conflict of interest	8
Change the order game	9
Cover pricing, withdrawal of tender	10
Compensation of tendering cost	11

Rank no. 1 = Most frequent. Rank no.11 = Least frequent

the whole industry. Liu et al. explained, however, that this notion of achieving consistency is linked to the different cultures which exist within the construction industry and this may blur the boundaries between ethical and non-ethical behaviour at times. A practical example of this could include the boundary between receiving a seasonal gift as a polite gesture and what is deemed to constitute an act of bribery to influence the award of a contract. Vee and Skitmore (2003) attempted to address this potential grey area and offered clarity on the boundary between gifts and bribery. They concluded that gift-giving transfers become an illegal act of bribery when they compromise relationships between the gift giver and the receiver and favour the interests of the gift giver. This is an important aspect for construction professionals, especially at the tender stages when bidders may offer them gifts or invitations to corporate functions, to gain a competitive advantage over their competitors. It is normal for construction professionals to have to sign anti-bribery legislation and declare any gifts to avoid accusations of impropriety in such cases. Other forms of unethical behaviour could include breaches of confidence, conflict of interest, fraudulent practices, deceit and trickery and also in some cases there may be problems of grey areas and interpretation difficulties. Moreover, less obvious forms of unethical behaviour could include presenting unrealistic promises, exaggerating one's expertise, concealing design and construction errors or over-charging (ibid.).

2.10 The need for construction professionals to uphold ethical and cultural values when procuring projects

> Have the courage to say no. Have the courage to face the truth. Do the right thing because it is right. These are the magic keys to living your life with integrity.

In consideration of the above quotation, the next important factor that arguably has a strong influence and link to professional ethics is culture and the cultural values of the construction industry and those construction client organisations that work within it. In the past, adversarial attitudes in the construction industry have affected relationships, behaviour, culture and trust. A major contributor to improving cultures within the construction industry has been professional ethics, which defines rules of conduct. However, construction professionals should regard their scope and ethical responsibilities as much greater and more extensive than just simply concerning conduct, institutional rules and regulations (Walker, 2009).

There have been opposing views on how best construction professionals can instigate cultural change within the industry and differentiation and fragmentation can again pose difficulties in this regard in aligning cultures, beliefs and standards (Liu et al., 2004). A practical example of this could be subcontractors having completely different standards, beliefs and values to the main contractors and similar scenarios existing between surveyors and architects. This raises the issue of the importance of changing the culture of the industry as a whole and the way it works. If cultural change is required, then a further question is raised, how do construction professionals working with their project teams achieve this? The answer could be in improved training, education and personal development to raise the awareness of the importance of professional ethics. Ahrens (2004) certainly supported this view and advocated the use of modules designed to teach ethics to built environment undergraduates to expand their knowledge and understanding of ethical issues affecting the construction industry, with particular emphasis on contracting responsibilities and liabilities. He explained that too many young practitioners graduating from higher education do not possess the skills in areas relating to ethical values, moral working, cultural difference and environmental responsibility and this is facilitating modules across 15 European universities to attempt to address this educational imbalance. Lui et al. (2004) presented a similar argument to improve ethical self-regulation and cultural change through education and training. Further supporting views came from Vee and Skitmore (2003), who explained that ethical codes alone are not sufficient to maintain ethical conduct. Their findings indicated that employer-led training and institutional continuous professional development (CPD) to educate members on what ethical codes mean from a practical perspective can greatly increase awareness and participation in ethical practice.

2.11 Conclusion

This chapter has presented the alternative descriptions of what ethics are which highlights that the differentiation and variation around a definition give rise to problems of ambiguity and meaning. Accordingly, an ethical dilemma ensues from the absence of any universally accepted definition of ethics. Perhaps this is one of the problems when considering professional ethics in the construction industry on a global scale, possibly leading to varying interpretations of ethics for construction professionals and what they mean in practice.

The goals of professional ethics can broadly be classified into two main categories: inward- and outward-facing goals. Inward-facing goals include creating common rules for people and organisations and outlining their responsibilities and duties to act and behave ethically. Conversely, outward-facing goals include providing a platform for the evaluation of professions and as a basis for public expectations.

The importance of ethics should not be underestimated as they are considered a vital and essential practice requirement, engendering the trust of the general public and preserving the employers' interests. Accordingly, one of the benefits of ethics for organisations revolves around trust, especially where firms are heavily reliant on their reputation in conducting business and gaining new work. Furthermore, construction professionals who act in an ethical manner will enhance their performance which will increase the success of the projects. Conversely, any compromises on ethics could jeopardise service delivery and damage the public view and the image of the construction industry.

It is important for construction professionals to commit to and encourage project teams to comply with sustainable ethical principles but there are many different aspects and issues that influence and affect professional ethics, and this presents a challenge for the industry. The codes of ethics, which have been introduced, have provided an indicator that organisations and institutions take ethical principles seriously, as they outline the expectations for all personnel with regard to ethical behaviour and intolerance of unethical practices. Notwithstanding this premise, unethical behaviour and practices have been experienced in the construction industry for many years. Repercussions can arise from non-ethical practices, especially in the context of the UK construction industry. Construction professionals, as leaders in the procurement of projects, should be leading the way in cultural change to improve the reputation of the industry. In this pursuit, they should be aware of the importance of ethics, its alternative definitions and the various interpretations of ethics, the reputation of the construction industry, codes of conduct and governance and regulations to avoid bad practices. In conclusion, it would appear that measures to improve the practice of professional ethics, such as professional codes of conduct, have gone some way to improve how the industry works but there are still far too many cases emerging of unethical practices blighting the sector. Although arguably these practices are due to a small minority of the sector, they create a bad press for the whole industry and further measures should be instigated by construction professionals to address this problem. Traditional responses in the past, at an institutional level, have been based on governance, regulations and punishment for non-compliance and clearly these have had only limited success. Perhaps construction professionals should be leading the way for a cultural change in the industry to train, educate and motivate construction individuals and organisations in what professional ethics entail, proposing measures to ensure compliance and the benefits that they can bring for the sector. This could be achieved through more focus on further education and higher education course modules linked to professional ethics and CPD through workshops and training events in the workplace. These measures will hopefully contribute to providing a more ethical environment for the industry and reap great benefits not just for all construction-related organisations and the building projects that they procure, but for the future of the construction industry at large. It is accepted, however, that to bring about these cultural changes will take conviction, integrity and, in some cases, courage not to engage in established unethical practices. These improvements once ingrained within the industry could then reap massive rewards in providing a safer, honest, trusting and more enjoyable working environment for all.

References

Abdul Rahman, H., Karim, S., Danuri, M., Berawi, M., and Wen, Y. (2007). Does professional ethics affect construction quality? Available at: www.sciencedirect.com (accessed 24 November 2019).

Abdul-Rahman, H., Wang, C. and Yap, X.W. (2010). How professional ethics impact construction quality: Perception and evidence in a fast-developing economy. *Scientific Research and Essays*, 5(23): 3742–3749.

Ahrens, C. (2004). Ethics in the built environment: A challenge for European universities. In *ASEE Annual Conference Proceeding*, pp. 5281–5289.

Chang, C.M. (2005). Challenges for the New Millennium. In *Engineering Management*. Upper Saddle River, NJ: Pearson Educational.

CMI (Chartered Management Institute). (2010). Code of professional management practice. Available at: www.managers.org.uk/code/view-code-conduct (accessed 24 November 2019).

CMI (Chartered Management Institute). (2013). *Codes of Ethics Checklist*. London: Chartered Management Institute.

Cowton, C. and Crisp, R. (1998). *Business Ethics: Perspective on the Practice of Theory*. Oxford: Oxford University Press.

Fewings, P. (2009). *Ethics for the Built Environment*. London: Routledge.

Guttmann, D. (2006). *Ethics in Social Work: A Context of Caring*. Philadelphia, PA: The Haworth Press.

Helgadottir, H. (2008). The ethical dimension of project management. *International Journal of Project Management*, 26, 743–748.

Internet Encyclopaedia of Philosophy. (2010). Ethics and self-deception. Available at: www.iep.utm.edu (accessed 24 November 2019).

Lere, J.C. and Gaumnitz, B.R. (2003). The impact of codes of ethics on decision making: Some insights from information economics. *Journal of Business Ethics*, 48(4): 365–379.

Liu, A.M.M., Fellows, R. and Nag, J. (2004). Surveyors' perspectives on ethics in organisational culture. *Engineering, Construction and Architectural Management*, 11(6), 438–449.

Poon, J. (2003). Professional ethics for surveyors and construction project performance: What we need to know. *The RICS Foundation in Association with University of Wolverhampton*, 148(9), 232.

RICS (Royal Institution of Chartered Surveyors). (2000). *Guidance Notes on Professional Ethics*. London. RICS Professional Ethics Working Party.

RICS (Royal Institution of Chartered Surveyors). (2010). *Maintaining Professional and Ethical Standards*. London: RICS.

Robson, C. (2000). Ethics: A design responsibility. *Civil Engineering*, 70(1), 66–67.

Rogers, R.A.P. (1911). *Short History of Ethics*. London: Macmillan.

Rosenthal, R. and Rosnow, R.L. (1991). *Essentials of Behavioral Research Methods and Data Analysis*, 2nd edn. Boston: McGraw-Hill.

Stansbury, C.S. (2008). *Examples of Corruption in Infrastructure*. London: Global Infrastructure Anti-Corruption Centre.

Trevino, L.K. and Nelson, K.A. (2004). *Managing Business Ethics*, 3rd edn. Hoboken, NJ: John Wiley and Sons Inc.

Uff, J.P. (2003). Duties at the legal fringe: Ethics in construction law. *Society of Construction Law*. Available at: www.scl.org.uk (accessed 24 November 2019).

Vee, C. and Skitmore, M. (2003). Professional ethics in the construction industry. *Engineering, Construction and Architectural Management*, 10(2), 117–127.

Vitell, C. and Festervard, D. (1987). Business ethics: Conflicts, practices and beliefs of industrial executives. *Journal of Business Ethics*, 6, 111–122.

Walker, A. (2009). *Project Management in Construction*. Oxford: Blackwell.

Weiss, J.W. (2003). *Business Ethics: A Stakeholder and Issues Management Approach*, 3rd edn. Columbus, OH: Thomson South-Western.

Wikipedia (2019). Ethics. Available at: www.en.m.wikipedia.org (accessed 16 July 2019).

3 The UK construction industry and ethical considerations

3.1 Introduction

This chapter will articulate and discuss the importance and significance of the construction industry to global economies including the UK, while highlighting the complex nature of the sector. Such complexities around the bespoke nature of building projects will be covered, alongside the temporary organisational aspects of the project procurement. Risks associated with the dynamic and fragmented environment that encapsulates the construction industry will be identified.

This chapter will also explain the roles and responsibilities of different members of project teams, including clients, project managers, architects, quantity surveyors, engineers, main contractors and subcontractors. It will consider these different roles from the perspective of ethical considerations and how they vary, depending on the individuals. As a consequence of these differing perspectives, dilemmas around ethical decision making will be discussed. Ethical investigation models will be identified which seek to address such dilemmas and assist construction practitioners in deciding what is right and wrong. Finally, the notion of building life cycles will be forensically analysed and the competing dilemmas of capital costs versus operational costs uncovered.

3.2 Context of the UK construction industry

The construction sector employs between 2–10 per cent of the total workforce in most countries of the world (Rahman and Kumaraswamy, 2004). In the UK, construction output in September 2020 accounted for more than £110 billion per annum and contributed 7 per cent of gross domestic product (GDP). Approximately a quarter of construction output is public sector and the rest is private sector. The construction industry is very diverse and ranges from the construction of buildings and infrastructure, maintenance, refurbishment to the manufacture and supply of building products and components. New build construction output accounts for approximately 60 per cent of construction output, whereas refurbishment and maintenance account for the remaining 40 per cent. The UK Government Construction Strategy states that there are three main sectors within the

UK construction industry: commercial and social (45 per cent), residential (40 per cent) and infrastructure (15 per cent).

According to the Government's Construction 2025 strategy (HM Government, 2013), the UK construction industry generates approximately three million jobs, which account for approximately 10 per cent of total employment and these include construction-related services and manufacturing. These jobs can be categorised as contracting, which employs approximately two million people, and is made up of 234,000 businesses and services, employing 580,00 people and 30,000 businesses. Furthermore, construction-related products account for 310,000 jobs and 18,000 businesses. The construction industry across the world can be regarded as a long-term industry linked to investment, albeit perceived to be a high cost, high-risk industry by some observers. The performance of the industry can be seen to be a good indicator of the state of the health of the wider economy. Accordingly, when an economy falters, it is common for the construction industry to slow down and when an economy begins to recover from a recession, the construction industry is one of the first industries to start to grow.

3.3 The bespoke nature of the construction industry

It is interesting to explore the origins of the construction industry as a bespoke project-based industry and there are many different theories on this. As a starting point, it is important to understand what makes the construction industry different from and potentially 'at odds' with most other industries. In answering this question, it is worth contemplating that the procurement process in construction is unlike that of most other industries. Those employed in construction are made up of mostly small teams, ranging from construction workers to design consultants, who come together on a temporary basis for the life of a project and then disband to undertake different projects. This creates fragmentation and does not always allow the time for relationships to develop and flourish, which could be in itself a contributory challenge for trust generation. It is also important to reflect upon the 'end product' and remember that construction projects are nearly always bespoke to clients' requirements. This 'one-off' or 'made-to-measure' aspect does, however, create risk and uncertainty for all parties. To fully appreciate and understand this context, we can compare the procurement of a new building with the purchase of a new car. When one buys a car, the make and model that suit your budget will be agreed, alongside any affordable optional extras that are required. One can even 'test drive' the same model to ensure that it meets your expectations in terms of feel and drivability. At this stage, when ordering the vehicle, you know exactly what you will receive on the due delivery date, which is normally a few weeks at most and you have an agreed fixed price. In this regard, there is very little risk that you will not receive exactly what you expected when you ordered, for the agreed price. As the car is made in a factory, it will be standardised, and quality control is normally very good. Completely the opposite scenario prevails when a building is being procured. It normally involves a prolonged period of time for design consultants to formulate a brief with clients, to progress the design development and tender the projects to construction contractors. On receipt of tenders, this is where the process probably varies most from the car purchase example. In selecting the most appropriate tender, it is important to consider the quality of the tenders rather than just accept the lowest price. Such factors as reputation, track record, resources and demonstration of an understanding of the project are vitally important, as the quality, cost, scope and timescale in delivering the final project are normally anything but assured. There are three elements which are widely regarded as critical success factors, namely cost, time and quality, and these constitute what is commonly referred to as the 'iron triangle', illustrated in Figure 3.1.

There are several unknown factors in any construction process which could cause the cost of projects to increase, programmes to be delayed and the quality of buildings to be compromised. This introduces two aspects which come into play around commerciality and risk and identifying who incurs any additional costs is frequently an area where disputes arise between clients and their consultants and contractors. Furthermore, given that construction works normally cost significant amounts of money, the stakes are high in terms of the final bill for clients and the level of profit attained by contractors. It is therefore perhaps not surprising, for reasons of commerciality, that parties to construction contracts have traditionally not relied on trust in dealings with each other, especially in financial construction matters.

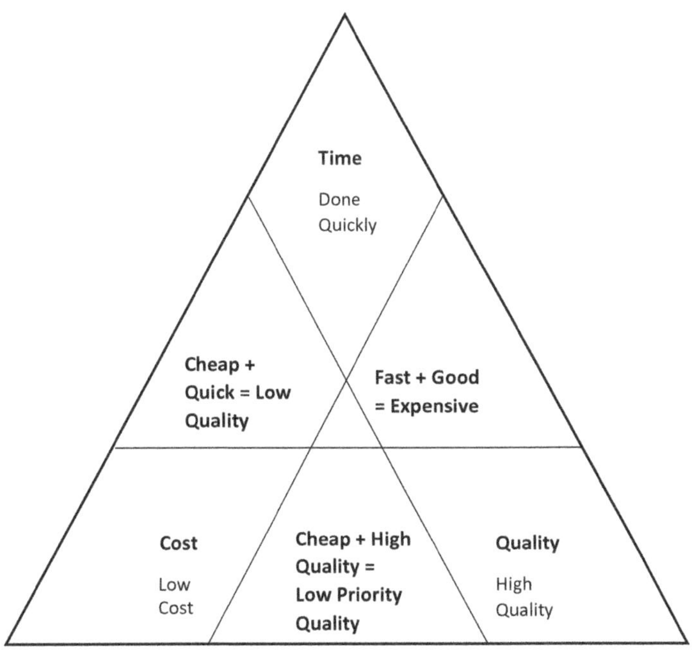

Figure 3.1 The iron triangle of cost, time, quality and scope

3.4 The fragmented nature of the construction industry

Despite the scale and importance of the UK construction industry, it is frequently criticised for performing poorly. In this regard, it has been accused of being adversarial, wasteful and dominated by single disciples reluctant to innovate and poor at disseminating knowledge. It is also claimed that the industry is fragmented and, according to Fewings (2009), there is a general tendency for those employed in the construction industry to ignore best practice, owing to this fragmentation within the sector. In the UK, this is reflected in the fact that nearly half of all work completed is undertaken by the majority of approximately 190,000 small contractors. Some of the workforce in this regard, especially those who undertake smaller jobs, are often one-off individuals with limited and potentially insufficient experience. As a consequence of such inexperience, there may be cases where their knowledge and awareness of issues relating to regulations, compliance, best practice and ethical considerations are lacking. In the context of the construction industry, which is associated with fragmented, complex and potentially confrontational practices, this creates a dilemma for the sector. It is, therefore, perhaps not surprising that there have been many reported cases in the press of examples of malpractice and contractors breaching regulatory standards and codes of practice. Some of these cases have resulted in litigation proceedings brought by their clients which has resulted in reputational damage for the industry as a whole. In addition, these practices have frequently led to less than acceptable project outcomes in terms of value for money, delays and poor-quality build standards.

3.5 The role of construction professionals in managing construction projects

In the organisational structure associated with managing construction projects, construction professionals can be employed by clients directly as their employees and they are commonly referred to as 'client-side representatives'. Alternatively, they can be appointed by clients as consultants to manage the design and construction of building projects on their behalf. In this scenario they normally would form a client design team, which could consist of consultant project managers, architects, quantity surveyors, mechanical and electrical engineers, structural engineers and other specialist consultants (Inuwa et al., 2015). It is normally the project managers and the overall design team appointed and led by clients who are responsible for developing the requirements of their clients, setting project briefs and managing the overall construction processes. Clients also employ their main contractors directly, and the main contractors under a traditionally procured building project will employ subcontractors. It is normally the subcontractors who appoint their respective suppliers and specialist sub-subcontractors.

The various aforementioned roles and responsibilities of construction professionals are detailed below.

3.5.1 Construction clients

Clients normally represent the sponsors of the project and therein have overall responsibility for the cost of the construction works and most of the risks involved. As they will be the 'end users' of buildings that are procured, it is important for them to ensure that buildings are designed and constructed to meet their detailed and prescriptive brief and functional performance requirements. In this regard, clients need to determine and set down the goals for their projects and the means by which they intend to deliver and achieve them. In this pursuit it is vital that they appoint the right specialist consultants who are experienced and offer expertise in the nature of a particular project. It is also imperative that they ensure that the right balance of resources is deployed in other areas and this includes budgetary and time considerations, to deliver projects safely and to an acceptable quality. For this reason, clients need to be equipped with the tools to lead their projects, which calls for informed and methodical approaches with clear responsibilities, roles and decision making.

Having clients at the forefront of projects with all the necessary skills in leadership, knowledge and resources has been highlighted as one of the main ingredients in improving the overall performance of the construction industry (Challender and Whitaker, 2019). Furthermore, maintaining the 'client voice' in decision making and identifying the brief have become more important in recent years and also not simply delegating these responsibilities to others in the professional team. This proactive leadership in the construction process, albeit challenging especially for 'lay construction clients', is essential in avoiding poor construction outcomes associated with time, cost and quality implications. In recent years, the calls for transformational change expressed in the Farmer Review of the UK Construction Labour Market (Farmer, 2016) and the Construction Sector Deal recognise the vital role of clients in striving to 'turn the tide' and spearhead such changes. Despite this, the construction industry has been slow to embrace strategies linked to client leadership. Accordingly, the unique role of clients and their preparedness in projects are emerging as a 'hot' topic. This is not sufficiently covered by the professional institutions where the focus is on the development of industry professionals rather than clients. Previous research within construction projects has mainly revolved around the development of professional teams, which is well-trodden ground and has increasingly diminishing returns on risk reduction.

The construction industry is a risk/reward venture undertaken by clients alongside their appointed project teams. Notwithstanding this premise, it is clients who ultimately take the biggest risk and if they represent a typical small business, arguably possess the least knowledge of protocols and culture of the construction industry (Challender and Whitaker, 2019). Within the structure of a construction project, a series of government-sponsored reports *Constructing the Team* (Latham, 1994), *Rethinking Construction* (Egan, 1998) and *Accelerating Change* (Egan, 2002) have made radical changes to the construction industry, making it more client-focused than ever before. A greater sense of teamwork and integration between clients and the design and construction supply chain has now significantly reduced design and construction risk to, and from, the client. However, from the point of view of designers and constructors, the client can still be seen as a 'risk'.

Part of this risk can be characterised as the dynamic shift that a novice client has to undertake to become a 'developer'. The client is required not only to take on the role of delivering their own business and operational change but also choose the right resources and create the right environment to successfully and seamlessly deliver a construction project. This is an enormous risk and very time-consuming for the corporate business, putting great strain on its management and physical resources. For many, it is a 'leap in the dark', therefore, the construction industry has few parallels in manufacturing.

3.5.2 Construction project managers

These construction professionals are normally regarded as having the overall responsibility for projects and accordingly are referred to as the lead consultant. They are responsible for all stages of projects from conception through to completion and sometimes are involved in the 'in use' occupation phase also. Their role includes allocating resources for projects and managing the other design team consultants to ensure that successful outcomes are achieved. This is especially important, considering the time, cost and quality aspects of projects. Accordingly, construction project manager responsibilities are arguably what guides a project to success. Construction project managers tend to have both good technical skills and leadership skills in directing their teams. In the latter case, communication skills are paramount in motivating a large project team and coordinating different contributions from many individuals.

Construction project managers work closely with architects, engineers and quantity surveyors to develop plans, establish programmes with timelines and milestones, and determine budgets for buildings. In developing these plans, they are ultimately responsible for ensuring that projects are completed on time, on budget and meet all the requirements of the client brief. Their role also includes reporting to their clients on the progress of projects at different stages and managing relationships with key stakeholders and external bodies.

3.5.3 Architects

These construction professionals are trained and qualified in the science and art of building design and normally, in the UK, are members of the Royal Institute of British Architects (RIBA). Their role is to develop the concepts for buildings and structures and translate these concepts into plans and images. Architects are responsible for creating the overall aesthetic and appearance of buildings and other structures, while ensuring that they are correctly designed to perform well against a predetermined client brief and its specifications. In this regard, they must ensure that buildings ae functional, safe and economical and meet the requirements of the end users. They need to consider how buildings will be used and the types of activity that will be conducted within them, while meeting all statutory requirements, including building and fire regulations.

Architects provide various designs and present drawings, illustrations and a report to their clients based on the project brief. Computer-aided design (CAD) and Building Information Modelling (BIM) technology have for many years replaced traditional drawing as the most common methods for creating design and construction drawings. The role of the architect is not completed at the design stages but continues on through the construction phases, where they have a responsibility to oversee and supervise the works. In the construction phases they are frequently requested to enact changes to the drawings and specifications in response to their client's changing requirements. Other variations may be predicated upon unforeseen site circumstances and physical constraints not envisaged at the design stages. These may include budgetary considerations or unforeseen ground conditions. As construction proceeds, architects must ensure that the construction contractors adhere to the agreed designs and specifications and use the correct materials and products. They are also expected to check and monitor adherence of the ongoing works from a construction workmanship quality perspective and ensure that the project programme is maintained.

3.5.4 Structural engineers

Structural engineers in the UK are normally members of the Institution of Structural Engineers. They design and oversee all structural aspects of building and infrastructure projects. This includes undertaking structural calculations, considering proposed loading and forces on building elements. This involves them in the preparation of structural design drawings and reports to ensure that buildings are completed to a safe and compliant standard and avoid damage or collapse when loaded. Alongside buildings, other structures can include retaining walls, bridges, foundations, structural frames and roads.

Structural engineering is related to the research, planning, design, construction, inspection, monitoring, maintenance, refurbishment and demolition of permanent and temporary structures. It also considers the economic, technical, aesthetical, environmental and social aspects of structures. As the architectural design progresses, structural engineers are required to design the main structural elements of buildings which include the foundations, floors and walls. Structural engineers make decisions on the type and quality of materials used for structural elements and this could include designing concrete to meet the correct loading requirements or strength. In this regard they determine the thickness, span and depth of structural members in the building, such as beams, and the size, quality, quantity of type of concrete reinforcements.

Structural engineers' designs are based on building codes which vary between countries and regions of the world and normally are regulated by national building regulations. Such regulations are normally predicated on the premise that under the worst load the structures will be subjected to in their lifespans, they must remain safe. During the construction phase, structural engineers regularly inspect materials and supervise the construction of structural members. They are required to approve completed works and this may sometimes involve testing materials. An example could be cube crushing tests to ascertain a required concrete strength has been achieved. In the event of building failures or collapse, they are brought in to carry out investigations and determine the causes, effects and solutions against reoccurrence.

3.5.5 Quantity surveyors

A quantity surveyor is a construction industry professional who specialises in estimating the value of construction works and managing all aspects of cost control on projects. They are normally members of the Royal Institution of Chartered Surveyors (RICS) and the projects they are involved in include new buildings, renovations or maintenance work. Furthermore, they can be employed on a wide variety of projects covering all aspects of construction, such as civil engineering, mining and infrastructure projects to determine the cost of such facilities. They are normally responsible for calculating early design costs for budgetary purposes through composite rates. When more design information becomes available, they can provide more accurate budget estimates using elemental cost methodologies. Quantity surveyors work closely with the other design team members and assist in the tender process. They produce tender documents, including detailed pricing schedules known as a bill of quantities, and manage the tender process, evaluating bid submissions alongside the other design team consultants.

The term quantity surveyor is derived from the role taken in quantifying the various resources that it takes to construct a given project, such as labour, supervision, plant and materials. They estimate and analyse the effects of design changes on budgets and deal with most of the contractual issues that arise on projects, including claims for variations and disputes. This involves overseeing elements of contract administration, clarifying and evaluating tenders, controlling variations, assessing contractual claims and negotiating and agreeing final accounts. They also are responsible for valuing the completed work and arranging for payments through a certificate of payment contractual process and sometimes they act as expert witnesses in litigation cases.

3.5.6 Building services engineers

Building services engineers are responsible for designing and overseeing all the mechanical and electrical building services installations on construction projects, including lighting, heating, gas installations and all electrical systems. They provide all the building services to give buildings functionality and which provide a stable internal environment to ensure that they have the correct temperature, air quality and lighting levels. This requires the provision of all the necessary backup support systems, such as power, hot and cold water and lifts. In addition, the installation of life protection systems, such as fire alarms and sprinkler systems, is an important responsibility of building services engineers. These functions should ideally be linked to sophisticated building management systems to ensure effective control and to minimise energy consumption.

In the UK, they are normally members of the Chartered Institution of Building Services Engineers (CIBSE). Building services engineers work very closely with the other design team consultants to ensure that their mechanical and electrical proposals are aligned and coordinated with the architectural and structural designs for a given building. They influence the architecture of a building and play a significant role in the sustainability and energy demands of building. Managing energy has become more important in recent years with the need to reduce carbon generation and make buildings more environmentally sustainable through use of renewable energy. Finally, building services engineers oversee the testing and certification of mechanical and electrical installation towards the completion of projects to ensure that they meet regulatory requirements, comply with the agreed performance specifications and conform to their designs. This is to safeguard that building services installations are fit for purpose and have not been compromised by the mechanical and electrical specialist subcontractors.

3.5.7 Main contractors

A main contractor, sometimes referred to as a general contractor, or prime contractor is responsible for the day-to-day oversight of a construction site, the management of vendors and trades, and the communication of information to all involved parties throughout the course of a building project.

It is the responsibility of the main contractor to execute all construction work activities that are required for the completion of projects. Their role takes on board many different aspects of construction management, including project planning, managing all health and safety aspects, overseeing legal issues, coordinating and supervising all construction activities and undertaking contract administration duties. They also appoint and manage all the construction subcontractors and ensure that the sequencing of all subcontracted packages of work activities are closely and carefully coordinated. In this regard, the main contractor brings a team of all the required professionals together, overseeing the construction while ensuring that all necessary measures are taken to execute a project effectively.

The main contractor is responsible for the preparation of plans and programmes to carry out construction projects. This ranges from hiring construction workers to developing a step-by-step timeline that the project will follow from start to finish. They are also responsible for hiring, supervising, and the payment of subcontractors and suppliers, alongside obtaining materials for the project to meet the specifications. Their other duties include acquiring all the necessary licences and permits from the local authority or relevant body so that building projects can progress legally and safely. Table 3.1 presents many other roles and responsibilities that main contractors undertake in the course of manging construction projects.

Table 3.1 The role and responsibilities of main contractors

Project planning responsibilities

- Planning important project development and implementation stages in advance of construction work commencing
- Managing various issues on projects, such as the requirement for materials and equipment
- Anticipating any potential modifications on projects
- Safeguarding that health and safety specifications are followed at all times
- Practising excellent communication between all parties involved in the construction process, such as clients and subcontractors
- Determining and addressing legal and regulatory requirements

Project management

- Managing the budget for the completion of construction activities
- Having responsibility to find and appoint the right subcontractors and individuals suitable for the nature of a given project
- Managing relationships with subcontractors and coordinating the equipment, materials and other services required by them for the duration of the project
- Submitting detailed valuations for payment of works completed

Project monitoring

- Monitoring of projects in terms of project programmes and key milestones
- Monitoring the quality of work carried out and checking adherence to detailed specifications and specialist design drawings
- Reviewing, modifying and updating project programmes and risk assessments to reflect the current status of projects, taking account of any variations in the scope of the works
- Implementing buildability measures throughout construction contracts, deploying value engineering methodologies and therein offering cost-effective building solutions

Legal and regulatory responsibilities

- Obtaining all necessary building permissions and permits
- Ensuring that projects are carried out in compliance with legal and regulatory issues

Health and safety responsibilities

- Monitoring all health and safety issues
- Creating viable and workable safety policies to ensure health and safety awareness is practised and maintained at all times. This may entail risk management strategies, emergency response systems and other preventative measures to ensure site safety.
- Ensuring that all construction site staff use safety equipment in the course of their work and comply with method statements and risk assessments
- Providing safety awareness for all staff

3.5.8 Subcontractors

According to NRM1 (RICS, 2012), the term subcontractor means 'a contractor who undertakes specific work within a construction project'. Normally subcontractors are specialists in a particular area or trade. They will perform all or part of the obligations of the main contractor's contract. In this regard, subcontractors have similar responsibilities to the main contractors in that they have to adhere to policies and procedures on site, provide personal protective equipment for their construction staff, ensure that the equipment they use is safe, and report safety hazards and incidents. They need to have regular communication with the main contractor and report to the main contractor's project manager on the progress of their work package. Any issues or problems in their work area must be communicated to the main contractor without delay to avoid any knock-on delay implications on the overall construction programme. This is largely owing to the dependency that one subcontract work package can have on another, owing to the sequential nature of construction work. This is especially important when a work package is on a 'critical path' and any delay to its completion will prevent another work package commencing.

Normally a large construction project or renovation often involves many different subcontractors who work together to complete projects in a timely manner. The nature of the subcontract works will determine the timing and sequence of when it is undertaken on site. For instance, a groundwork or piling subcontractor will often be one of the first subcontractors to commence and complete their work. Conversely, a decorating or flooring contractor may be one of the last subcontractors in the sequence, as their work is planned at the end of the contract programme. Depending on the subcontract agreement, they may be responsible for providing their own materials and equipment for the work package they are employed to undertake.

3.6 Different perspectives on ethics

The meaning of ethics can vary depending on the different roles and responsibilities of the many various individuals who make up project teams. For instance, ethics, as viewed from a client's perspective, could revolve around a project team which is honest, truthful and presents accurate and transparent progress reports and information at all times. Having a project team that can be trusted is therefore of paramount important for clients who will ultimately be the sponsors of the project and responsible for the cost of the construction works. Clients will also expect ethical behaviour from their teams in managing projects closely and reporting any potential risks. Clients need to be made aware of any potential issues on projects especially related to cost, time and quality matters. If their project teams do not keep them fully appraised of issues that can potentially cause them difficulties, then this could be regarded as unethical behaviour and is likely to lead to a loss of trust. In situations where unexpected incidents on projects arise, which have not been predicted by those responsible for managing, this can lead to a loss of confidence from the client's perspective. Similarly, if members of the project team are perceived to charge clients monies which are deemed to be unreasonable and not what they agreed or expected, then once again they could result in a loss of confidence. Another common source of frustration for clients can emanate from contractors not dealing with defects and snagging post contractor. There is a contractual and ethical responsibility for contractors to address reported building issues in a responsible and timely way, and delays which can cause clients disruption can cause tensions in the client-contractor relationships.

From the consultant project team's perspective, there may be other ethical considerations than those of their clients. These may include expectations that clients are transparent in keeping them abreast of all issues especially around budgets and any issues that may affect the progress of construction projects. For instance, if clients require approval from third party stakeholders, e.g. funders, then some of the project team should be aware of dates in this regard so that these milestones can be accommodated into the programme timelines. The other ethical elements from the project consultants' perspectives is to be reimbursed for any additional work which they are instructed to undertake. What can sometimes be frustrating and damaging for client-consultant relationships, according to Fewings (2009), is where clients as paymasters expect their consultant to undertake additional scope of services at no cost.

From the perspectives of the main contractors, subcontractors and suppliers, their view on ethics can be different again from those of clients and consultants. One of their main concerns is to be paid on time. Timely payments for these parties are especially important as cash flow for them is critical in their financial dealings. A lot of construction companies are forced into administration as a direct result of delayed payments to them. Accordingly, contractors and suppliers are sensitive to any instances where their employers pay them outside timescales which they have contracted to. This practice is regarded as unethical and could lead to disputes and, in some cases, legal action.

In addressing some of these ethical issues from the consultants' and clients' perspectives, Challender et al. (2014) stated that building collaborative working relationships between the project teams, including the wider supply chain, is essential. A 'collaboration toolkit' was devised, with intervention measures to promote trusting relationships and partnership working. This was designed to manage problematic issues on projects in a proactive and responsive way before they progress to potential disputes between parties.

3.7 Decision making from an ethical standpoint

The construction industry, as previously noted, is dominated by many different organisations and individuals from a whole range of diverse backgrounds employed in their roles to make decisions of behalf of others. There is an implied duty of care alongside established values and standards which govern whether decision making can be deemed good, bad or indifferent. This can help inform whether decisions are right or wrong in a situation or context. In theory, it would be easy to surmise that it is clear and obvious in most cases to know what represents poor decision making. However, in practice, the picture is not always black and white but different shades of grey, brought about by subjectivity rather than objectivity. This can create problems for the construction industry in deciding how we judge what is right and wrong and therein what is acceptable. To assist in this dilemma, Harrison (2005) created a model for ethical investigation predicated on different branches of ethics. These are summarised in Table 3.2 and can assist in deciding whether a particular situation or decision is ethical or unethical.

In consideration of the above, in reality, more than one approach may be used and it really depends on the context in the situation being investigated.

According to Fewings (2009), there are different models relating to ethics in decision making. The 'moral intensity' model is related to the moral impact on decisions by reflecting on the extent of the consequences that might ensue from certain decisions. It would also seek to assess and consider the social backlash from certain options and the probability that negative outcomes could result from those decisions. In such circumstances, it would try to predict which stakeholders or members of the community could be affected and to what degree. Some would argue that considering only the negative outcome of decisions is unethical in itself as this could be construed as a 'damage limitation' exercise and not derived from ethical and moral judgements to do the right thing and adopt the most ethical pathway.

The 'business model' (ibid.) is a commercially driven approach based on maximising financial outcomes and values. This is predicated on maximising profits but also considers whether approaches are ethical and legal in this process. For this reason, it is important for an organisation using this model that the legal and ethical considerations are not outweighed by the maximisation of profits. We have seen many different examples in the past, especially in the financial services sector, where financial considerations have dominated the

Table 3.2 A model for ethical investigation

Ethical category of investigation	*Description*
Descriptive ethics	Uses experiential learning and empirical studies to aid and direct comprehension within a certain context. Gives precedents and principles for guidance for decision making.
Normative ethics	May provide one answer dependent on a particular view that is adhered to
Philosophical ethics	Provides a moral answer by defining the dilemma but in a very theoretical way
Practical ethics	Provides a 'moral compass' as a guide to acceptability. Provides practical models and rules of application to suit organisational values

modus operandi of companies and where a culture of non-compliance with legal and ethical standards has emerged. The global recession of 2009 was allegedly brought about by irresponsible and unethical merchant bankers and the downfall of old-established banks such as Merrill Lynch on Wall Street. There are many lessons to be learnt from such cases and with this approach it is important to have closely managed, supervised, robust procedures for adherence to regulatory and ethical standards to avoid a repeat of such scandals. Such procedures and processes should be built on a system of doing the right thing and should have transparency, integrity, impartiality and respect at their core.

A 'virtuous or professional model' is predicated on a strong focus on decision making for the 'greater good' and the benefit of society. It is built on strong moral principles in upholding cultural and social norms and looking for ways and means of continuous improvement (ibid.). An example is environmental ethics, which is concerned with reducing carbon generation, tackling the climate emergency, promoting good health, reducing waste (especially plastics) and promoting more sustainable communities. Education clearly plays a major role in promoting such an approach and environmental and sustainability teaching and learning programmes in schools and universities are now common in the developed world. In a construction industry context, there has been a recent push to consider the ways and means to reduce the waste embodied in building processes and use more environmentally sustainable materials and products. In addition, there has been a more concerted and integrated consideration of 'life cycle' costs rather than simply capital costs in the feasibility stages of projects. This is a welcome move and considers the renewable technologies and energy-saving measures which could reduce the 'carbon footprint' of a building over its complete lifespan. The virtuous model also covers 'social and corporate responsibility' of organisations. This relates to companies doing the right thing for communities and the public at large, and having responsible policies on such aspects as employment, working conditions, health and safety, pollution and contributing to local and national agendas.

An 'integrity and reputational model' is derived from the premise that organisations that create and build a strong ethical reputation over many years benefit from the public confidence that this bestows on them and their brand. When organisations adopt this model, it is not always for altruistic reasons but is linked to improving their image or creating positive public relations to increase their market share. An example in the UK could be Marks and Spencer, who for decades have maintained a reputation for providing good quality products at a reasonable price. This integrity and reputational model can generate increased sales or work orders for organisations and potentially increase profits in this way.

3.8 The life cycle of buildings

The man-made environment in which we live is sometimes referred to as the 'built environment'. In considering the built environment, one needs to consider this from the perspective of the life cycle of building and engineering structures. The life cycle in this regard does not look only at the build period, known commonly as the construction contract period or construction phase, but extends to the operational/occupation period of buildings, known as the 'in use phase' (Fewings, 2009). Prior to the construction period, part of this notion of life cycle includes the development phases, involving the inception, planning and tender stages of projects. Furthermore, the decommissioning and subsequent demolition of buildings and engineering structures can be regarded as the final phases of the life cycle, which may transform into the life cycle of a new building.

During the life cycle of buildings there are many different organisations and individuals involved with differing roles and responsibilities and these were articulated and discussed earlier in the chapter. There are also competing interests at large, especially at the design and construction stages. Decision making at the early stages of the life cycle can have long-lasting effects on later phases such as the 'in use' occupation phase. It has long been accepted that the cost of a building's life cycle during the occupation, commonly referred to as operational costs or 'opex costs', normally represents approximately 80 percent of the total cost of a building over its life. Conversely, the costs during the development and construction phase, commonly referred to as capital costs, or 'capex costs', normally represent only 20 percent of the total. Despite this, it is normally only the initial capital costs of the build that are commonly considered in business cases as part of the decision-making process at the early financial viability stages. This can have negative connotations for projects whereby they could be compromised by too much emphasis on the initial construction costs. As an example, it is not uncommon for decisions to be made at these early stages to reduce the capital costs through 'value engineering' of the building design which could have long-term implications for the future performance, maintenance and running costs of the buildings. An example of this dilemma is demonstrated in Box 3.1.

Box 3.1 Example of early decision making affecting the long-term future of buildings

A client organisation has plans to redevelop their accommodation by procuring a new building. As part of this process they develop a strong business case for the new build, predicated on a feasibility study prepared by their appointed design consultants. In the feasibility study the total projected costs for the building were estimated as £20 million, based on an elemental cost breakdown prepared by the appointed quantity surveyor acting as the cost consultant. During the next stages of design development, when the full implications of the detailed design are being considered, and progressed, through consultation with end users, it becomes apparent that the build is more complex and sophisticated than what was initially envisaged. The anticipated project costs at this later stage reflect this added complexity and are revised to £25 million.

This increase in capital costs now creates a question for the client organisation around the financial viability of the increased value of the project, since the anticipated business outcomes remain the same. A value engineering process is undertaken with the client and design team assessing where cost savings can be made on the build. A 'shopping list' of potentially desirable rather than essential scope of works is considered as part of this rationalisation process. Decisions are made to redesign certain elements of the build for more cost-effective solutions and in some cases certain items of work are omitted. The principal considerations at this stage are purely to 'balance the books' and make sufficient capital cost savings to make the scheme viable. There is little if any foresight or consideration of how these modifications and omissions will affect the performance, maintenance and running of the new facility.

Other considerations linked to sustainability are simply discounted on the basis of affordability. The decision making at this critical juncture over the life of the building has severe implications in restricting the performance of the facility and making it unable to fulfil those initial anticipated outcomes articulated in the business plan. In addition, the costs of running the facility over the 30-year life of the building have increased, owing to the omission of renewable energy solutions, increasing gas and electricity consumption.

Maintenance costs have also increased significantly, as a result of less robust and arguably more inferior materials, components and building systems being installed. The cumulative effect of savings £5 million on the capital costs and bringing the project 'back on budget' has in fact added an additional £40 million over the 30-year life of the building in operational costs. This impact is further compounded by a view that the functional capability and capacity of the building have been compromised to the extent that it no longer performs the operational improvements initially envisaged.

The above scenario hopefully demonstrates that early design making not only can have long-lasting consequences on the performance of buildings but also can impose overarching restrictions on organisational operations, with potentially severe implications for businesses. It also demonstrates that short-term planning can have long-lasting detrimental financial effects which could far outweigh the original cost savings. It is worth considering this scenario in the context of professional ethics as those decisions made by relatively few individuals at the design and construction phases can potentially affect many others in the occupational stages. There is also the notion that those making the decisions at the early stages to keep the capital costs within budget often compromise the build, knowing full well that such decisions will be to the detriment of the build in the long term. Some would argue that these individuals could be compromised in the cost-cutting process and that this is unethical and immoral behaviour on their part. Others could suggest that pursuing strategies linked primarily to meeting capital affordability budgets, at the expense of other aspects such as sustainability and specifically carbon generation, is short-sighted and misguided from an ethical standpoint.

In consideration of the above points, there is a long-standing view that the construction industry is predicated on meeting short-term needs rather than focusing on long-term sustainable solutions. Accordingly, it has been criticised for its general lack of longer-term future visioning for projects, with insufficient emphasis at the decision-making stages on full life cycle considerations.

3.9 Summary

The construction industry accounts for a large percentage of the global economy and employs between 2–10 per cent of the total workforce in most countries. The performance of the industry can be seen to be a good indicator of the state and health of the wider economy. Despite the scale and importance of the UK construction industry, it is frequently criticised for performing poorly. This could be attributed to a general tendency for those employed in the construction industry to ignore best practice and be resistant to change.

The construction industry has emerged as a bespoke project-based industry which has many different characteristics from other industries, such as manufacturing. This is largely due to construction projects nearly always being unique and 'tailor-made' to suit clients' individual requirements. This bespoke project-based industry is made up of mostly small teams, ranging from construction workers to design consultants. Such teams come together on a temporary basis for the life of a project and then disband and this makes it very different from most other industries. This also introduces fragmentation which presents an ever-increasing challenge for the sector. Owing to the fragmentation and bespoke nature of its composition, there are several unknown factors in any construction process which could cause the cost of projects to increase, programmes to be delayed, and the quality of buildings to be compromised. This creates two aspects which come into play around commerciality and risk and who incurs any additional costs.

Project teams are made up of many construction professionals who have different roles and responsibilities. Clients are normally the project sponsors and accordingly they should be instrumental in steering, resourcing and leading all stages of the construction process from concept to completion. They appoint design teams, which normally consist of project managers, architects, structural engineers, mechanical and electrical engineers and quantity surveyors. All these professional consultants work together closely within their different specialist areas during the design and construction phases. This coordinated and collaborative approach is designed to enhance teamwork in procuring successful project outcomes. On the contracting side of project team, there are main contractors, subcontractors and suppliers. Main contractors are responsible for the day-to-day oversight of construction site activities, which includes management of vendors and trades, and the communication of information to all involved parties throughout the course of the building project. Normally subcontractors are specialists in a particular area or trade, and they will perform all or part of the obligations of the main contractor's contract. It is common for a large construction project or renovation to involve many different subcontractors who work together to complete projects in a safe and timely manner.

The meaning of ethics can vary depending on the different roles and responsibilities of the many various individuals who make up project teams. Clients, for instance, will expect ethical behaviour from their teams in managing projects closely and reporting any potential risks to them, as issues can frequently create additional costs and time delay implications. Conversely, from the consultant project team's perspective, there may be other ethical considerations to those of their clients. These may include expectations that clients are transparent in keeping them abreast of all issues, especially around budgets and any issues that may affect the progress of construction projects. Ethics can mean something different to main contractors, subcontractors and suppliers. One of their main concerns is to be paid on time and to be treated fairly in commercial and contracting matters. Timely payments for these parties are especially important as cash flow is critical in their financial dealings.

There is an implied duty of care alongside established values and standards which govern whether decision making by construction professionals can be deemed good, bad or indifferent. However, there are sometimes problems for the construction industry in deciding how to judge what is right and wrong and therein what is acceptable. Ethical models predicated on different branches of ethics can assist in deciding whether a particular situation or decision is ethical or unethical. These include the 'moral intensity', 'business', 'virtuous or professional' and 'integrity and reputational' models.

The life cycle of a building does not consider only the build period, known commonly as the construction contract period or construction phase, but extends to the operational/occupation period also, known as the 'in use phase'. There are competing interests at large between the design and construction phase and the in-use phase, normally predicated on capital costs versus operational costs. Decision making at the early stages of the life cycle can have long-lasting effects on the occupation in-use phase. Operational costs during the in-use phase normally represent most of the total costs of a building over its life whereas the capital costs during the development and construction phase represent only a fraction of total costs. Despite this, projects can be compromised by too much emphasis on the initial construction costs where capital savings may be required to meet budgetary requirements. These savings are sometimes referred to as 'value engineering' and this short-term planning can have long-lasting detrimental financial effects, which could far outweigh the original costs saving.

Taking the different roles, responsibilities and perspectives that this chapter has discussed, Chapter 4 will articulate ethical dilemmas which could be faced by construction professional in the course of carrying out their duties.

References

Challender, J. (2017). Trust in collaborative construction procurement strategies. In *Proceedings of the Institution of Civil Engineers: Management Procurement and Law*. London: Institution of Civil Engineers.

Challender, J., Farrell, P. and Sherratt, F. (2014). Partnering in practice: An analysis of collaboration and trust. *Proceedings of ICE: Management, Procurement and Law*, 167(6): 255–264.

Challender, J. and Whitaker, R. (2019). *The Client Role in Successful Construction Projects*. Abingdon: Routledge.

Egan, J. (1998). *Rethinking Construction: The Report of the Construction Task Force*. London: TSO, pp. 18–20.

Egan, J. (2002). *Accelerating Change: Rethinking Construction*. London: Strategic Forum for Construction.

Farmer, M. (2016). *Modernise or Die: The Farmer Review of the UK Construction Labour Market*. London: Construction Leadership Council.

Fewings, P. (2009). *Ethics for the Built Environment*. London: Routledge.

Harrison, M.R. (2005). *An Introduction to Business and Management Ethics*. Basingstoke: Palgrave Macmillan.

HM Government (2013). *Construction 2025. Industry Strategy: Government and Industry in Partnership*. London: HM Government, pp. 23–25, 61–71.

Inuwa, I.I., Usman, N.D. and Dantong, J.S.D. (2015). The effects of unethical professional practice on construction projects performance in Nigeria. *African Journal of Applied Research (AJAR)*, 1(1): 72–88.

Latham, M. (1994). *Constructing the Team*. London: The Stationery Office.

Rahman, M.M. and Kumaraswamy, M.M. (2004). Contracting relationship trends and transitions. *Journal of Management in Engineering*, 20(4): 147–161.

RICS (Royal Institution of Chartered Surveyors). (2012). *NRM1, New Rules of Measurement*. London: RICS.

4 Corporate social responsibility, social value and ethics

4.1 Introduction

This chapter introduces corporate social responsibility (CSR), social value (SV) and ethics in the context of the construction industry. It starts by discussing the history and evolution of CSR and SV, understanding what drove these concepts forward and how they are underpinned by ethics and ethical decision making. This chapter also describes how the construction industry engages with ethics at an organisational level, and concludes by illustrating how CSR and SV are manifestations of ethical practice and are used as evidence of ethical compliance and behaviour, somewhat regardless of their organisational motivations and origins. A discussion on the repositioning of traditional 'profit first' arguments is also included, as is the influence organisational actions have upon the actions of employees, further validating the wide-ranging benefits ethical behaviour has for all stakeholders.

4.2 Defining CSR

Defining CSR could be a book in itself. However, for the purpose of this chapter we need only reach an agreed understanding as to what CSR entails. There are fundamentally two distinct approaches that can be taken. The first is to have a definition so precise that it only meets the interpretations of a single organisation. While this will describe the organisation's CSR activities perfectly, it will be problematic to then reach agreement with external parties. The second approach is to understand CSR as a general idea, an umbrella term under which sit a wealth of interpretations. This will allow different stakeholders to reach a broad agreement as to what CSR means. It will, however, then prove difficult to reach an exact agreement as different stakeholders inevitably have different perceptions and interpretations of CSR. For the purposes of this chapter we are proceeding with the latter and will outline the broad concept of CSR and the different ideas that sit within it, as they apply to the construction industry.

One proposed definition that was widely adopted and served as a foundation of the CSR debates was introduced by Carroll in 1979, and later updated in 1983 and 1991 (Carroll, 1991). This was known as the 'Pyramid of corporate social responsibility' (Figure 4.1).

Figure 4.1 Pyramid of corporate social responsibility
Source: Adapted from Carroll (1991).

Figure 4.1 shows the steps a company will go through in their CSR journey. Carroll argued that companies will at first ensure they are able to make money (fulfilling their economic responsibility). After this has been achieved, a company will then focus on their legal responsibilities and ensure they abide by all required legislation. The third step companies will take is to then fulfil all ethical responsibilities. Only then will a company undertake any philanthropic activities. Carroll defines CSR as the fulfilment of all levels of his pyramid. Therefore, we can conclude that Carroll broadly defines CSR as a company's positive economic, legal, ethical and philanthropic activities.

Discussion 4.1

Do you think Carroll's (1991) definition of CSR is correct? Do you agree with the structure of what an organisation aims to achieve first, i.e., profit before abiding by legislation?
What would this mean for the motives of companies in general? And is this the same across all industries?

Wood (1991) has contributed much to the discussion on CSR over the years and takes the stance that CSR is when a company increases activities that benefit society and reduces any behaviours that harm society. While this definition is wide-reaching and can encompass many company strategies and practices, it starts from the idea that companies are fundamentally in the wrong and their operations bring a degree of harm to society. Others have taken a more positive approach. Zhao et al. (2012) consider CSR to be how well a company links its economic, social and environmental policies to the interests of stakeholders and the needs and values of wider society. Murray and Dainty (2009), in one of the first textbooks to tackle the concept of CSR from a construction industry perspective, argued that actions must be voluntary (an idea we return to later in this chapter) and decisions must be taken in a socially ethical manner with consideration for how they impact on society.

This chapter defines CSR as the ethical strategies and practices a company undertakes to address social, economic and environmental problems in society, independent of whether they are caused by the company.

Discussion 4.2

Do you think organisational ethics is the same as CSR?
By adopting CSR principles in some areas of their operations, does this mean a company is behaving ethically?

4.3 A brief history of CSR

CSR is often considered a modern phenomenon, but its origins can be traced back hundreds and perhaps even thousands of years. We won't be tracing the origins that far back, instead we will focus on a few select examples that have furthered the CSR cause and helped shape how we consider CSR today.

Born in 1836, Joseph Rowntree was a businessman and philanthropist. Upon hearing his name, most people may think of sweets. However, Joseph Rowntree used much of his wealth to set up charitable trusts with the aims of positively impacting housing and living standards and the quality of life not only for his workers but also for the wider society. This was not called CSR at the time, but the actions of the Rowntree Company fulfil modern-day CSR ideas. Similarly, the actions of the Cadbury brothers (of Cadbury's chocolate) in the late 1800s included building an entire village to house their workers to provide cleaner and safer living conditions. While again the term CSR itself was not used then, the behaviour exhibited by Cadbury fulfils this criteria. However, such actions introduce a very interesting ethical question regarding the seemingly responsible actions of companies.

Discussion 4.3

If a business benefits from its own 'CSR', does this still count as CSR?
Do a company's actions need to be selfless to be considered ethical?

What Rowntree and Cadbury did for society in the 1800s was unquestionably of benefit when it came to improving living standards across several criteria; from cleaner, more spacious housing, to increasing education through the building of schools and providing fresh countryside locations for people to live and work. However, it must also be acknowledged that such actions allowed their workforce to be fitter and healthier (to work longer and harder) and have a happier home life (and so suffer from reduced amounts of distractions while at work). The social benefit Rowntree and Cadbury brought to the 1800s (and still do to this very day) is unquestionable, and I am sure all workers would rather be happier and healthier, even if this does mean they will be working longer and harder. However, from an ethical standpoint, can such actions be considered to come under the banner of CSR, or does a company need to engage in practices from which they will not benefit, in order for any actions to be considered truly philanthropic?

Discussion 4.4

Are there any CSR practices a company can engage in that they will not ultimately benefit from in an indirect way?

The recent emergence of the idea of CSR came in 1953 when Howard Bowen published a book entitled *Social Responsibilities of the Businessman* (Bowen, 1953). This book is credited with pushing the theme and ideas of CSR into the mainstream consciousness of wider society. Bowen was arguing that the problems in society resulting from the Second World War should be addressed by industrialists and those businessmen who had prospered during the war or had successful existing businesses. This call for increased business responsibility started to raise the expectations society held for business. Such increased expectations then continued to grow (albeit slowly) over the proceeding decades, until a large shift to incorporate environmental responsibility occurred in the 1960s.

An American marine biologist called Rachel Carson played a large part in furthering the responsibilities of organisations with her book *Silent Spring* (Carson, 1962). *Silent Spring* focused on highlighting the negative impacts that pesticides had upon the ecosystem and the wider environment. While this is not immediately and directly related to the construction industry, the release of *Silent Spring*, and the accompanying publicity, led to an increased critical view of business by society. Individuals were no longer willing to accept the transgressions of companies simply to further industrial progress and wealth of society in general. The publication of *Silent Spring* inevitably led to a backlash from some industrial groups who tried to minimise its significance and scientific robustness. However, the book was based on extensive research and verified by leading scientific experts, and so the link was firmly established in the minds of individuals; pollutants and pesticides made by industry negatively impacted our ecosystems. And so began a drive for company accountability that ultimately contributed to the environmental movement entering mainstream consciousness.

The environmental movement was then sustained during the 1970s with an increasing CSR focus also emerging on how companies were treating local communities and if they were being good neighbours (Elbert and Parker,

1973; Carroll, 1999). This was to be achieved through both minimising the harm they caused by making changes to their daily operations and maximising the positive benefits they could achieve. It was during this decade the performance of companies was monitored against both the CSR goals they set themselves and the CSR expectations of society.

The 1980s saw a decline in the CSR behaviour of companies, which perhaps somewhat sceptically, could be linked to the downturns and recessions in the early and late 1980s.

Discussion 4.5

In times of economic hardship (downturns and recessions), if a company scales back its CSR practices, does this mean such practices were never truly ethical and only conducted in times of economic prosperity? How committed can a company be to CSR practices if they drop them so easily?

If a company is experiencing financial difficulties, should they temporarily stop engaging with expensive CSR practices in order to focus on keeping the business afloat and ensuring the workforce remains in employment?

The 1990s brought with them a wave of change for both politics and business, with increased prosperity and wealth generation occurring throughout the UK and beyond. Accompanying this increase in wealth was a stronger expectation for companies to engage in CSR, a momentum which was sustained and built upon as the millennium approached and passed. The first two decades of the millennium saw a sustained change in the expectations placed upon business in general. The reporting of organisational CSR behaviours is undertaken annually by many businesses, and some companies are more likely to have an advertising campaign highlighting their CSR than they are on the products and services they sell. The concept of CSR has also expanded to include the full range of a company's reach, including adherence to modern-day slavery regulations, child labour restrictions, working hours, carbon emissions, diversity of staff and health and safety record. Such a list of practices is not exhaustive, but the expectations are now that the CSR requirements of a company will also apply to the company's supply chain. Stakeholders expect a company to ensure all the other companies it does business with operate to the same high standards and expectations.

4.4 CSR as a form of business ethics

Ethics has been discussed in relation to the development and definition of CSR, but can a company's ethical practices and CSR initiatives be considered to be one and the same? Can an unethical company engage with CSR as a ruse to obfuscate its unethical business operations? Or is a company's ethical practice manifested in its CSR?

Discussion 4.6

Is a cigarette company, that manufactures and sells a product that is linked to causing death and disease, but gives a large portion of their profits to charitable causes as part of their CSR, be considered ethical?
Can a company be considered ethical if the very essence of its operations causes harm to society, even if through its profits it tackles other societal problems?

If we take Carroll's (1991) ethics definition of an 'obligation to do what is right, just and fair' and to 'avoid harm', then arguably a company whose very operations cause some form of societal harm cannot be considered ethical. The true picture, however, is rarely ever that simple. Construction as an industry, for example, heavily exploits natural resources, and then causes further disruption in the use and application of these resources. However, without the construction industry, we would not have the very buildings in which we work, the houses where we live and the infrastructure to get from one to the other safely and effectively. Therefore, a certain degree of disruption, and even harm could be considered acceptable if the operations of the company result in a greater positive benefit for society. While cigarette companies may not fall into this category, arguably construction companies will do.

If a company operates in an ethical manner in its business transactions and provides a benefit to society through the products it delivers, then CSR could ultimately be considered an extension to a company's ethical practice. CSR is also often a point of reference for companies when they are experiencing any adverse publicity. They attempt to deflect any negative stories by pointing to the 'benefits' they have contributed to society through their CSR practices, in the hope these will deflect public attention away from the bad publicity. Research does show that if a company engages with new CSR practices shortly after a widely publicised transgression has occurred, then it will be viewed with a high degree of scepticism by the general public. Long-standing CSR practices, however, do allow a certain amount of forgiveness from the general public with transgressions judged more sympathetically (Watts et al., 2016).

Discussion 4.7

If a construction company engages in CSR practises as expected by society, but does nothing else, does this mean they are an ethical company?

The public do perceive a company's ethics and CSR differently; however, research has shown that almost three-quarters of the public believe a company's CSR and a company's ethics are of equal importance. The same research concluded that customers value a company's CSR higher than the company's ethical behaviour when forming opinions of that company. However, it was reported that ethics have more of an impact upon the public's perceptions of a company's brand than CSR. This was due to ethical breaches by a company being perceived to directly harm potential customers, whereas CSR breaches lead to 'only' indirect harm. Customers have no trust in companies that display poor ethical choices, however, where companies are perceived to behave ethically, this can overcome any shortcomings resulting from actions constituting negative CSR. Similarly, CSR actions can help overcome any negative ethical issues, but only in relation to improving the company's brand. Overall, the research concluded a company's ethics are perceived as separate from CSR, with ethics having more of an impact on the perceived company brand by customers than CSR does (Ferrell et al., 2019).

We can therefore conclude that CSR and ethics are perceived as different constructs but with a highly complex and interdependent relationship. A relationship so interdependent that many often confuse the two and see a company's CSR practices as evidence of its ethical behaviour. While CSR strategies and practices cannot remove the need for a company to behave ethically, such practices do go a long way to improving the perception of the company with potential customers when ethical questions are not answered satisfactorily. There are numerous other benefits to CSR practices that companies can experience that are explored in more detail in Section 4.5.

4.5 The advantages (and disadvantages) of CSR

The advantages of CSR can be wide-ranging, and while the positive impacts upon the intended recipients can appear obvious, historically, there has been great difficulty in evidencing the benefits of a company's CSR in any meaningful and easily communicable way. Great efforts have been made by both industry and academia to measure the positive impacts CSR practices bring (and the negative impacts such practices alleviate), all in an attempt to evidence the advantages of CSR. There is good reason to this – how better to evidence ethical behaviour by companies reporting on their CSR practices than to be able to show the difference your practices have made in firmly measured numbers and figures? Initially, when the main societal focus of CSR was upon the environment, companies started to measure and discuss both the damage they were preventing (i.e. a certain amount of carbon offset or a reduction in the use of a certain type of material) and also the positive impacts they were responsible for (i.e. the planting of trees). As the concept of CSR evolved to focus more on social responsibility, the advantages still needed to be evidenced. This is where measurement turned to the number of apprenticeships employed and the number of work experience weeks offered. Financial metrics are also heavily used to illustrate the benefits company CSR practices bring and include figures such as the total amount of local spend and the amount of money the public sector has saved per individual now in employment.

Research has shown that there are also numerous benefits to be experienced by a company by engaging in CSR, including an increased reputation among stakeholders that can enable companies to distinguish themselves from competitors in a positive way (Carroll and Shabana, 2010). CSR engagement can also ultimately lead to more consumers switching to purchase the company's products (Du, Bhattacharya and Sen, 2010). Increasing CSR will also help a company become an employer of choice by increasing their appeal among potential future employees (Backhaus, Stone and Heiner, 2002). This appeal will also extend to the perceptions of the company by all staff, as research has also shown that by embracing CSR, a company will benefit from higher-performing and innovative employees who exhibit increased levels of commitment, with the company experiencing lower amounts of staff turnover and higher levels of staff satisfied with management (Gaudencio et al., 2014).

Discussion 4.8

If a company gains a financial benefit, either directly or indirectly, from engaging in CSR, does this mean their motives are driven by economic gain rather than by ethical orientation?

Is there anything incorrect or wrong about a company engaging with CSR practices for the sole purpose of increasing their financial performance?

It is perhaps somewhat illustrative of our current society that companies who are perceived to be behaving ethically receive a competitive advantage for doing so. The starting point therefore must be the perception that companies behave unethically as part of their normal business practices. If the behaviour of all business was considered ethical, there would be no distinguishing factor between the different companies that operate within the industry. This is perhaps where CSR practices help evidence a company's true ethical commitment, in that, by engaging with CSR practices, the company is making a statement of intent to go above and beyond the ethical standards its competitors achieve, and the ethical standards that are expected by society. Members of society can evidence their displeasure at a company's unethical actions by engaging in a range of responses from not purchasing goods to protests directed at the company themselves. However, 'minimum' standards do exist that a company must abide by and achieve, in the form of legislation, which in effect forces companies to follow some form of basic ethical standards. It is therefore the deviation from such standards, or the failure to exceed them, that will ultimately lead to poor reputational perceptions within society.

4.6 CSR and ethics as a form of governance

There are numerous methods by which the ethically desired behaviours of construction companies can be enforced and encouraged. Proactive techniques can include the inclusion of subcontractor requirements in construction tendering exercises. Reactive techniques can include the introduction of legislation.

Legislation can take two forms in the United Kingdom; primary and secondary. Primary legislation is created by Acts of Parliament and is therefore a time-consuming and expensive process but one which is enacted by representatives of the people, who have been voted in to create such legislation in a democratic manner. The word 'Act' is used to identify such legislation, for example, the Public Services (Social Value) Act (2012), which we will consider in more detail later in this chapter. Secondary legislation is created by a branch of government where the relevant powers have been delegated by Parliament. This could involve the amending of existing law or the creating of regulations by a branch of government that could contain specialists in the subject field. Such a method of legislation can arguably close loopholes discovered or taken advantage of in primary legislation and can be a quicker and cheaper process of law creation but is arguably not democratic, as those directly involved in the creation process may not be elected officials. The word 'Regulations' can be used to identify such statutory instruments, for example, the Building Regulations and the Scheme for Construction Contracts (England and Wales) Regulations.

The reason some examples of legislation can be described as reactive is that in most circumstances something occurs, either before legislative action is then taken to prevent it occurring again or to encourage it to occur more frequently. For example, construction has historically been an industry suffering high injury and fatality rates. To help combat this, many initiatives and pieces of legislation have been introduced over the years, such as in 1994 when the first iteration of the Construction (Design and Management) (CDM) Regulations was introduced. The CDM Regulations were then updated in 2007 and again in 2015 to take account of construction aspects not included in the earlier versions, and any technological and process changes that had occurred in the industry since the previous release.

When legislation is in place, regardless of how it was introduced, it governs the behaviours and responsibilities of both individuals and companies within its remit. Such governance through legislation can come to dictate standards of ethically acceptable behaviour. If a company does not abide by legal requirements, it is often seen as unethical in its operations. However, this then raises the question of whether ethics extend beyond what is legally required of a company. If we return to Carroll's pyramid of CSR in Figure 4.1, it argues that a company will ensure it is profitable in the first instance, and then abide by legislative requirements, before ensuring it is ethical, and then finally engaging in philanthropic practices. Ethics and legislation are therefore viewed as distinct and different phases of a company's journey. Before any legislative requirements are introduced, a company could theoretically behave in any manner it deems appropriate. As long as such behaviours are not breaking the law, they could continue indefinitely. In practice, however, this may not always be the case. It is not a legal requirement that company reports on their CSR, yet many do as it is now a somewhat expected behaviour in many industries. Conversely, it often takes legislation and other statutory instruments to ensure company compliance with initiatives that some would expect to be common practice. Take, for example, the Equality Act (2010) (Gender Pay Gap Information) Regulations (2017), which made it a requirement for companies with over 250 employees to publish their gender pay details. This requirement revealed that in many companies there was a disparity in what was paid to men and women. In some cases, men and women were paid differently for doing the same job, which is illegal. However, in other cases, the requirement to publish the pay details of their employees revealed a huge disparity between men and women generally, which is arguably unethical but not against any current laws. The construction industry, for example, had an average pay gap, between men and women, of 23.3 per cent in both 2017 and 2018. Since the forced publication of such data, most of the leading construction companies have launched initiates aimed at recruiting more females into the industry, as they currently make up around 12 per cent of employees and so this does go some way to explain the disparity in pay. Nevertheless, if the construction industry had addressed these issues previously, such legal requirements would not have been necessary – it is regrettably through governance that most companies seem to stand up and address key ethical issues. Only then do some try and go 'above and beyond' what is required.

Discussion 4.9

Is a company ethical if, before the introduction of the Equality Act (2010) (Gender Pay Gap Information) Regulations (2017), they paid their female and male employees differently?
Is a company ethical if they have a workforce comprising of 95 per cent males?

The discussion points in this chapter are largely for debate and to trigger deep consideration, there is often no clear-cut right or wrong answer, although sometimes you may arrive at a position you feel strongly to be correct. It is the very philosophical nature of ethics that others may have a completely contradictory belief that they equally feel is correct. In considering the second question in Discussion Point 4.9, it can be unethical to hire only one sort of employee, especially if this is a result of nepotism or cronyism. However, some types of job may naturally attract one sort of person, and there is no law against a workforce being 95 per cent male or female. Nevertheless, extensive studies have shown the benefits that a company can experience from having a diverse workforce (McCuiston et al., 2003; Podsiadlowski et al., 2013; Vardeman-Winter and Place, 2017). There is also a fairly recent phenomenon of positive discrimination to be aware of. This is where a company will seek to hire a certain profile of individual and then purposefully exclude any other kinds of individuals from the job. While this could be derived from good intentions, attempting to make their workforce more diverse, if it is considered an act of discrimination, it could therefore also be ultimately unethical, despite the advantages of having a more diverse workplace. Examples of overcoming such ethical quandaries include allowing any individual who applies for a job who is underrepresented in your current workforce to automatically be put through to interview. This is a well-used method that appears to work successfully for some employers.

The Public Services (Social Value) Act (2012) is another example of how the behaviours and practices of companies are governed by legislation. The Act compels public sector bodies engaged in procurement to consider the economic, social and environmental benefits tenders can bring, and to award contracts based on overall best value and not simply the most cost-effective one. The Social Value Act can be seen as a method by which public bodies can save money and take a more cohesive approach to the myriad of services they procure and provide. In practice, it allows a public body to ask any construction contractor what 'additional social value' they will be providing when deciding to award a project. Any 'additional social value' can then be considered when making the award decision.

This Act was raised by Chris White, an MP from 2010–2017, with the intention of giving small and medium-sized enterprises (SMEs) and social enterprises a greater opportunity to be successful in public sector procurement. The wider benefits that SMEs offer could be considered alongside the criteria of time, cost and quality. The Act also legitimised the public sector in allowing them to consider the most beneficial tender across a variety of criteria. The Act was first introduced as a Private Members Bill. A Private Members bill is where an MP who is not currently part of the Cabinet proposes a bill to be debated by the House of Commons, which then can ultimately be passed into law and become an Act of Parliament.

In the case of the construction industry, the Public Services (Social Value) Act (2012) is another example of governance over what could be described as company ethical practices. There was nothing stopping companies before the introduction of the Act from offering social value practices with all their returned project tenders. However, now in order to be competitive, construction companies must offer some sort of social value. This could be considered from two different perspectives. The first is that companies didn't offer any social value beforehand, and so the government felt compelled to ensure private sector companies used some of their vast resources in assisting public bodies solve any issues they were facing for no additional cost. This takes the view that construction companies do not conduct social value practices on their own and instead attempt to make the maximum profit from each contract, irrespective of the social issues they could help prevent or reduce. This is a somewhat pessimistic view, albeit one that you may consider to be true. The second perspective you could take is that before the introduction of the Social Value Act, some construction companies might have focused on generating social value with each project they tendered for, through, for example, the use of local labour and materials, the hiring of apprentices and donating to local charities. This could have resulted in higher overall costs being submitted.

Public bodies, which are under scrutiny for how they spend public money, before the introduction of the Act, perhaps would have been compelled to award projects to the most cost-effective contractor. This would effectively mean that all contractors who offered social value that resulted in their prices increasing would more often than not be unsuccessful in procurement opportunities. The introduction of the Social Value Act therefore legitimised public sector bodies to award projects on best overall value, therefore, giving weighting to social value practices. Arguably if this perspective is correct, it means the Social Value Act was not intended to govern poor company behaviour but instead was intended to enable public sector clients to have the power to reward those companies already engaging in good practices. The truth is probably somewhere in between these two perspectives, with the Public Services (Social Value) Act (2012) both allowing public bodies to reward companies who engage in social value practices and attempting to raise the minimum standards of all construction company social value behaviour through increased governance.

Discussion 4.10

Why do you think the Public Services (Social Value) Act (2012) was introduced?

4.7 The emergence of SV

The 'social' element has always been a key ingredient in the concept of CSR. Although, as discussed earlier in this chapter, this element historically took a back seat to make way for a more environmental focus. Such a focus was understandable, as during the past 50 years our impact upon the environment has become apparent. Widely publicised negative impacts have, therefore, been at the forefront of CSR practices, as they are accompanied by societal expectations for action. For example, when CO_2 targets were first discussed, any contractor who didn't mention their CO_2 reduction strategies somewhere would have been asked why and treated with a healthy dose of suspicion by potential clients and the wider general population. As some environmental targets could easily be measured and communicated, it was logical that these then became the focus of construction company intentions. Any progress against such goals could then be measured and reported to evidence a company's CSR. This went some way to satisfying the construction industry's stakeholders, who demanded action be taken to address their environmental concerns. Any company that took actions, and published strategies on dealing with environmental issues, was seen to be taking CSR, and by extension, ethics towards the environment, seriously. However, with increasing and changing societal expectations, the environmental actions companies undertook under the banner of CSR were almost taken for granted, and so the focus began to shift to other aspects of CSR, namely, social value (SV).

4.8 Defining social value

According to Social Value UK (2020b), social value can be described as 'the quantification of the relative importance that people place on the changes they experience in their lives'. The Social Value Portal (2020a) describes social value as

> an umbrella term for these broader effects, and organisations which make a conscious effort to ensure that these effects are positive can be seen as adding social value by contributing to the long-term wellbeing and resilience of individuals, communities and society in general.

The Public Services (Social Value) Act (2012) itself does not clearly define social value, but does state that in the context of procurement, social value should be considered as a proposal that can 'improve the economic, social and environmental well-being of the relevant area'.

It can be argued therefore that SV is the positive impacts felt by individuals and wider society as a result of organisational actions. This could be intended 'extra' actions undertaken by an organisation separate from their daily activities, for example, a construction subcontractor who donates profits to local charities and gives free career talks at schools and colleges. SV could also include the actions companies undertake as part of their operations, such as the hiring of apprentices or procuring goods and services from social enterprises. SV could also be generated from a business's core operations alone, such as the construction of a hospital, youth centre or the regeneration of an entire neighbourhood in need of modernisation.

It is also worth discussing that the evolution of SV is much like the evolution of CSR. As the concept of CSR changed over time due to some environmental concerns being addressed, other concerns increased in importance, and so did the changing requirements of clients and stakeholders. SV is following and will follow a similar trajectory. From the few definitions discussed, we can see SV is a nebulous and subjective concept, meaning different things to different people. Research has shown that stakeholders can hold different interpretations of CSR mainly due to the different requirements in the expectations (Watts et al., 2015). As these requirements and expectations change, so does the idea of what SV is and what it relates to. This can be problematic for construction professionals and companies who are faced with communicating and delivering SV. A recent study has shown that such difficulties are perpetuated not only by a difference in SV understanding between stakeholders, but also by the changing understanding of each constituent stakeholder. What is considered SV one year may not be considered SV the next (Watts et al., 2018). Therefore, any company trying to understand SV and reach a firm definition will need to be aware of the futility of such an approach, and the disadvantages faced with arriving at a 'fixed' definition of SV that they are unwilling to abandon.

Discussion 4.11

What are the potential disadvantages for a company of having a 'fixed' understanding (a definition that they do not change) of social value?

However, even though definitions and understandings of SV can evolve and differ, a shared idea of the fundamental principles that underpin SV can still be reached. Engaging with SV is ultimately about making a positive difference to the social fabric of society. While different understandings and ideas exist about how best to achieve this, behind any actions and requirements are arguably the same motivating factors; to make the world a better place by behaving, and encouraging others to behave, in an ethical manner.

4.9 SV, CSR, ethics and the construction industry

The construction industry encompasses the design, construction, operation, maintenance, refurbishment and demolition of buildings, engineering works, infrastructure projects and built assets. Therefore, more than any other industry, construction has a very 'real' presence in the everyday lives of all the population, from the houses we live in, powered by the energy services we need, such as electricity, gas, broadband and water, to the infrastructure we use to travel to work, to the offices, warehouses and factories, in which we spend our days. The construction industry plays a major role in shaping all of our lives. The industry itself also largely conducts its entire operations in the public eye. From road repairs witnessed every day, to redevelopment of city centres and the construction of houses, flats, hospitals, office blocks and car parks, to name but a few. It is an industry like no other, where public scrutiny is never-ending, and every stage of a construction project's life cycle is witnessed and commented upon. This makes the construction industry unlike any other in that any transgressions or the appearance of unethical behaviour will be highlighted, reported and commented on by a wide range of stakeholders almost immediately.

The industry, however, has also always suffered from a bad reputation, and has been perceived as one that has shown little regard for the heavy exploitation of natural resources or pollution of the environment, and evinces an attitude of indifference to wider society (Barthorpe, 2010). Construction has also always been an industry that has been viewed as lagging behind others when it came to addressing these issues and embracing the wider ethical principles of CSR (Glass, 2012). This is perhaps why it is has been reported that the construction industry has more reason than most to adopt CSR and SV principles, in an attempt to address its poor image and highlight some of the many positive ethical behaviours it exhibits and has exhibited for many years.

Indeed, CSR and SV may be relatively new concepts to the industry over the past few decades, but examples of behaviour evidencing high ethical principles can be witnessed from organisations operating within the construction industry going back centuries. However, the arguments outlining the construction industry's ethical transgressions persist, and arguably with some good reason. While it is unfair to judge all operatives and all businesses on the actions of only a few, it is these few examples that often grab the headlines and undermine the positive efforts being made by the vast majority of construction companies and professionals. Against such a negative reputation, those who operate in the industry need not only abide by ethical standards, but also must actively showcase their ethical compliance to fight against the tide of bad publicity. As it is, therefore, often difficult for a construction company to evidence it is behaving ethically, many companies heavily promote their CSR and SV behaviours through social media, annual reporting, wider marketing and in all tender returns.

4.10 Construction industry examples of ethical (and unethical) behaviour

There are many great examples of historic and contemporary construction industry ethical behaviours. As discussed in this chapter, companies operating within the construction industry largely evidence their ethical behaviours and compliance through their CSR and SV practices, but there are also examples of wider positive ethical behaviour that companies often engage in.

4.10.1 The Willmott Dixon Foundation

Established in 1852, Willmott Dixon are one of the largest privately-owned construction contractors operating in the UK. They have won two Queen's Awards for Enterprise, are regularly voted as one of the *Sunday Times*' Best Big Companies to work for and have twice won the 'Best Main Contractor to Work For' Construction Enquirer Award for contractors whose turnover is above £250m. Willmott Dixon have five core beliefs that support what they do. These consist of Sustainable, Excellence, Contributing, Quality and Ethical. The Ethical core belief is that their 'people and operations meet the highest standards of conduct' (Willmott Dixon, 2020).

In 2011, Willmott Dixon launched their flagship Willmott Dixon Foundation. This is a manifestation of their company values and their family-led approach and the ethical principles that guide them as an organisation. While undoubtedly good publicity and its use in tender returns allows the Foundation to be used to win more work, its very existence and continued support evidence the ethical approach to business adopted by Willmott Dixon. The Foundation is a great example of ethics in practice, with the aim of enhancing the life chances of 10,000 young people, a target which in 2019 they met one year earlier than they originally had set out to achieve.

4.10.2 WSP's compliance and ethics programme

With a history that stretches back over 130 years, WSP are an internationally recognised professional services organisation that has grown with the acquisition of numerous specialist companies in recent years. According to WSP, they 'develop creative, comprehensive and sustainable engineering solutions for a future where society can thrive' (WSP, 2020a). The WSP website has a quotation from their Chief Ethics and Compliance Officer that evidences their commitment to achieving the highest ethical standards in their business operations 'Our working relationships are based on trust, and putting the highest ethical standards at the centre of all we do …' (WSP, 2020b).

WSP lists numerous policies freely available on its website in an attempt to be fully transparent with all stakeholders in showcasing the approach they take to be an ethical organisation. All of these policies are collectively known as the 'Code' and govern the behaviour of all staff. There are many motivations for such a proactive and public approach to ethical standards and adherence, but one reason is that before clients and customers even engage with WSP, they are fully aware how important the company believes behaving ethically to be. The 'Code' is extensive and covers general codes of conduct, anti-corruption policies, gifts, entertainment and hospitality as well as insider trading and working with third parties.

4.10.3 The construction industry blacklisting scandal

Unfortunately, despite many construction companies reflecting the ethical policies and ethical practices used as examples here, there are also quite a few ethical transgressions apparent in the construction industry. As they are often largely publicised, they go a long way in undermining all the positive work done elsewhere in the industry, and although some are becoming quite dated, they still live strong in the recent memories of those directly and indirectly affected.

The construction 'blacklist' was a list that included the names of over 3,000 construction operatives. It was argued those on the list were denied employment by several important construction contractors over many years. It is believed the list was compiled with the purpose of identifying individuals who were trade union activists. However, individuals were added to the list without their knowledge, for reasons that were not disclosed and there was no system in place for reviewing or appealing. The scandal was revealed and, in 2015, several leading contractors issued a statement that contained several admissions of guilt. There was also several court hearings and a substantial compensation scheme set up to settle the scandal with over 256 individuals receiving compensation.

This practice was highly unethical and is one of the major unethical events that has occurred in the construction industry in recent times. It is especially infamous due to the sheer number of parties involved and the length of time the practice continued. However, this scandal does reveal that unethical practices can occur, and companies can move onwards, albeit by acknowledging their indiscretions, apologising and offering compensation. Whether it is felt this is a suitable form of apology or not, all but one of the contractors involved are still operating today, with the only one no longer operating: Carillion. Once considered one of the giants of the construction industry, it was not a direct unethical action, such as the blacklisting scandal, that led to Carillion entering liquidation in 2018. While investigations are still ongoing into the precise circumstances surrounding Carillion, it was inevitably a multitude of factors, perhaps some of them ethical, that led to their eventual collapse.

Discussion 4.12

Can a construction company ever make up for past unethical behaviour?
Does it depend on how bad the unethical behaviour was?
Does it depend on how big the apology gesture is?

When cases such as the blacklisting scandal are revealed, it is also easy to forget that although organisations receive bad publicity and it is the company name associated with the unethical behaviour, it is ultimately select individuals within those companies who are responsible for the ethical transgressions. This can be either directly through unethical acts, or indirectly through the promotion and tolerance of an unethical culture.

4.11 Ethical frameworks

One method by which organisations encourage ethical practice among construction employees is by supporting them to achieve a professional accreditation. Professional bodies exist to set and uphold ethical standards, as the Royal Institution of Chartered Surveyors (RICS) state: 'Behaving ethically is at the heart of what it means to be a professional; it distinguishes professionals from others in the marketplace' (RICS, 2020). The RICS go further, with the production of their five global professional and ethical standards. Adherence to such standards demonstrates an individual's (and, by extension, a company's) commitment to ethical behaviour. Table 4.1 provides the authors' opinions of how these ethical standards can be applied to the culture and practices of a construction organisation.

4.12 CSR, SV, ethics and the alternative argument

It is also worth including here the idea that organisations engaging with CSR and SV are still actually engaging in unethical acts. To understand this argument, we need to first understand some of the arguments put forward by the Nobel Prize-winning economist Milton Friedman. In a *New York Times*

Table 4.1 The RICS ethical standards

RICS ethical standard	*The authors' opinions on how the ethical standards apply to company practices*
Act with integrity	This standard is concerned with transparency and being honest. Companies should be straightforward in their dealings, with practices in place that are transparent and promote trust. They should not take advantage of clients and have clear policies in place that govern all circumstances in which the integrity of staff could be questioned, such as on gifts, hospitality and conflicts of interest, etc. CSR and SV actions are often undertaken and communicated to illustrate the integrity of organisations.
Always provide a high standard of service	Companies should strive to provide clarity and consistency in decisions. This includes acting within the competence, remit and scope of the abilities of staff by not taking on too much work or work that staff are not able to competently deliver in the timeframes required.
Act in a way that promotes trust in the profession	While the remit of companies does not extend into the private life of employees, companies should have clear expectations of how professionals should act both in and outside of the work environment. This should include policies on issues such as social media and other public profiles and forums. It is also important companies (and their staff) stay true and do what they say they will. One important characteristic of this standard is being aware of the spirit of any requirements, and not just the letter. In practice, this means behaving in the spirit of a contract, i.e. being fair to all parties, and not just using the contract against other parties in times of genuine confusion and difficulty.
Treat others with respect	A somewhat obvious standard but nevertheless a standard of key importance. This is paramount to a vibrant, trusting and enjoyable workplace environment and culture. The ethical behaviour of all staff in the workplace needs to be expected at all hierarchal levels, with support and guidance set out clearly to ensure everyone is aware of what is required of them in all situations.
Take responsibility	There needs to be a high degree of accountability in any company, where blame is not unfairly distributed, and staff feel empowered to take action to raise concerns and address issues they feel are important. Companies should have complaint handling procedures in place and transparent processes to evidence how all issues raised are taken seriously and treated fairly. CSR and SV practices are also undertaken with the purpose of showing wider stakeholders that, as a company, responsibility is being taken for addressing some societal concerns.

Magazine article published in September 1970 called 'The Social Responsibility of Business Is to Increase its Profits', Friedman argued that a company's responsibility is to its shareholders. This was expanded upon by saying that shareholders are the only group to which a company is socially responsible, and all employees of that company should work with that main purpose in mind (Friedman, 1970). Friedman was not arguing against the idea of ethical behaviour, but that a company should 'make as much money as possible' while ensuring they conform to the 'basic rules of society, both those embodied in law and those embodied in ethical custom'. In the article, Freidman clearly argues that contribution to any ethical or social cause above the minimum required by law is not acceptable as this will reduce the overall profits generated by the company.

If this argument were to be used by modern business, it would eliminate the need of many companies to engage with CSR and SV altogether. However, as both stakeholder and wider societal expectations have shifted in recent times, this has served to reposition the arguments proposed by Friedman. There may be no legal requirement for many of the practices companies undertake in the name of CSR and SV, but by not engaging with CSR or SV, companies would be leaving themselves exposed to criticism in the marketplace and in the eyes of stakeholders, and so may be less successful in tendering for new projects. Indeed, studies have shown that those companies who engage with CSR and SV are more likely to retain and recruit staff, enhance job satisfaction and increase financial performance (Brammer et al., 2007; Du et al., 2007; Saeidi et al., 2014).

4.13 Summary

This chapter has explored the history of CSR and SV, sought to explore how these concepts can be defined (and the difficulties associated with such a task), and also provided some case study examples of ethical practices in action and how a recognised ethical framework can be used by companies operating in the construction industry. Corporate social responsibility and social value can ultimately be viewed as an extension of an organisation's ethical practices. If individual ethics can be considered as the values that govern an individual's behaviour, then an organisation's ethics can equally be considered as those self-determined rules guiding the way the organisation operates. CSR can be argued to be the manifestation of the way in which the organisation operates. If a construction main contractor pays all subcontractors within 30 days, you will read about it in the main contractor's CSR report. If a construction consultancy allows its staff to take paid time off to work with good causes, you will again read about it in their CSR report. Arguably all CSR and SV activity can be viewed as an organisation's ethical actions in practice. It is how they illustrate to stakeholders and to the wider society they are ethical companies.

References

Backhaus, K., Stone, B. and Heiner, K. (2002). Exploring the relationship between corporate social performance and employer attractiveness. *Business & Society*, 41(3):.267–268.

Barthorpe, S. (2010). Implementing corporate social responsibility in the UK construction industry. *Property Management*, 28(1): 4–17.

Bowen, H. (1953). *Social Responsibilities of the Businessman*. New York: Harper & Row.

Brammer, S., Millington, A. and Rayton, B. (2007). The contribution of corporate social responsibility to organizational commitment. *International Journal of Human Resource Management*, 18(10): 1701–1719.

Carroll, A. (1991). The pyramid of corporate social responsibility: Toward the moral management of organizational stakeholders. *Business Horizons*, 4: 39–48.

Carroll, A. (1999). CSR: Evolution of a definitional construct. *Business & Society*, 38: 268–295.

Carroll, A. and Shabana, K. (2010). The business case for corporate social responsibility: A review of concepts, research and practice. *International Journal of Management Reviews*, 12(1): 85–105.

Carson, R. (1962). *Silent Spring*. Boston: Houghton Mifflin.

Du, S., Bhattacharya, C. and Sen, S. (2010). Reaping relational rewards from corporate social responsibility: The role of competitive positioning. *International Journal of Research in Marketing*, 24(3): 224–241.

Elbert, H. and Parker, R. (1973). The practice of business: The current status of corporate social responsibility. *Business Horizons*, 16(4): 5–14.

Ferrell, O., Harrison, D., Ferrell, L. and Hair, J. (2019). Business ethics, corporate social responsibility, and brand attitudes: An exploratory study. *Journal of Business Research*, 95: 491–501.

Friedman, M (1970). The social responsibility of business is to increase its profits. *The New York Times*, Magazine. Available at: http://umich.edu/~thecore/doc/Friedman.pdf.

Gaudencio, P., Coelho, A. and Ribeiro, N. (2014). Organisational CSR practices: Employees' perceptions and impact on individual performance. *International Journal of Innovation Management*, 18(4): 1–26.

Glass, J. (2012). The state of sustainability reporting in the construction sector. *Smart and Sustainable Built Environment*, 1(1): 87–104.

McCuiston, V., Wooldridge, B., and Pierce, C. (2003). Leading the diverse workforce: Profit, prospects and progress. *The Leadership and Organisational Journal*, 25(1): 73–92.

Murray, M. and Dainty, A. (eds) (2009). *Corporate Social Responsibility in the Construction Industry*. London: Routledge.

Podsiadlowski, A., Groschke, D., Kogler, M., Spring, C., and van der Zee, K. (2013). Managing a culturally diverse workforce: Diversity perspectives in organisations. *International Journal of Intercultural Relations*, 37: 159–175.

RICS (Royal Institution of Chartered Surveyors) (2020). The Global Professional and Ethical Standards. Available at: www.rics.org/globalassets/rics-website/media/upholding-professional-standards/standards-of-conduct/the-global-professional-and-ethical-standards.pdf.

Saeidi, S., Sofian, S., Saeidi, P., Saeidi, S.P. and Saeidi, S. (2014). How does corporate social responsibility contribute to firms' financial performance? The mediating role of competitive advantage, reputation, and customer satisfaction. *Journal of Business Research,* 68: 341–350.

Social Value Portal (2020a). What is Social Value? Available at: https://socialvalueportal.com/what-is-social-value/#:~:text=Social%20Value%20serves%20as%20an,communities%20and%20society%20in%20general.

Social Value UK (2020b). What is Social Value? Available at: www.socialvalueuk.org/what-is-social-value/#:~:text=Social%20value%20is%20the%20quantification,they%20experience%20in%20their%20lives.&text=Examples%20of%20social%20value%20might,next%20to%20a%20community%20park.

Varderman-Winter, J. and Place, K. (2017). Still a lily-white field of women: The state of workforce diversity in public relationships and research. *Public Relations Review*, 43(2): 326–336.

Watts, G., Dainty, A. and Fernie, S. (2015). Making sense of CSR in construction: Do contractor and client perceptions align? In A. Raiden, and E. Aboagye-Nimo (eds), *Proceedings of 31st Annual ARCOM Conference*, 7–9 September 2015, Lincoln: Association of Researchers in Construction Management.

Watts, G., Dainty, A.R.J. and Fernie, S. (2016). *The influence of public sector procurement practice in shaping construction CSR*. Paper presented at RICS Annual Construction, Building and Real Estate Research Conference (COBRA 2016), Toronto, 19–22 September.

Watts, G., Dainty, A.R.J. and Fernie, S. (2018). Paradox and legitimacy in construction: How CSR reports restrict CSR practice. *International Journal of Building Pathology and Adaptation*, 37(2): .231–246.

Willmott Dixon (2020). Company overview (2020). Available at: www.willmottdixon.co.uk/about-us/company-overview#:~:text=The%20core%20beliefs%20that%20support,the%20environment%20we%20all%20share.

Wood, D. (1991). Social issues in management: Theory and research in corporate social performance. *Journal of Management*, 17(2): 384–406.

WSP (2020a). What we do. Available at: www.wsp.com/en-GB/what-we-do.

WSP (2020b). Acting with integrity. Available at: www.wsp.com/en-GB/who-we-are/corporate-responsibility/ethics-integrity.

Zhao, Z., Zhao, X., Davidson, K. and Zuo, J. (2012). A corporate social responsibility indicator system for construction enterprises. *Journal of Cleaner Production*, 29–30: 277–289.

5 Social value, procurement and ethics

5.1 Introduction

This chapter summarises the concepts of procurement and tendering and applies them to the construction industry. The procurement practices of the construction industry are described, as are the evolving rules and regulations governing procurement behaviour and actions. The role of ethics in procurement is introduced and the importance of making ethically sound procurement choices is explored. The immediate and indirect ramifications and implications of incorrect ethical decisions are also considered and how such consequences can then be best addressed in an ethical and transparent manner is discussed. The focus of this chapter is on the increasing prevalence of ethics and ethical decision making in construction procurement and how existing ethical frameworks can be applied and understood in the context of procurement decisions. How ethical frameworks can be applied to different procurement methods, ethical supply chain management, and the part ethics plays in payment practices are all also discussed as these concepts are linked closely to, and in a number of cases are derived directly from, the procurement practices adopted. The benefits of ethical procurement are discussed throughout the chapter as behaving in an ethical way offers both short-term and long-term benefits for all stakeholders involved, as well as improving the image and productivity of the wider construction industry.

5.2 What is procurement and tendering?

Procurement is quite simply the act of obtaining goods or services. This is usually a transaction with the goods and/or services provided by party A to party B, and, in exchange, party B will pay an agreed sum of money to party A. The act of procurement is completed on a daily basis by most people without any serious thought – such as the purchase of a cup of coffee or a sandwich. This can be considered procurement in its simplest form. In the UK construction industry, procurement is a little more complex.

The RICS has published extensively on procurement within the construction arena, with the first edition of a Professional Guidance note entitled 'Tendering

Strategies', published in 2014, (RICS, 2014). It is important to note that RICS Guidance Notes are for guidance purposes only and are compiled by practitioners to provide examples of best practice and recommendations for industry.

The RICS Guidance Note 'Tendering Strategies' starts by attempting to clear up a common misconception within the UK construction industry between the definition of procurement and the definition of tendering. Procurement refers to the 'overall act of obtaining goods and services from external sources' In construction, this could include the strategies a main contractor enacts in producing and sending out work package enquiries and the comparison and awarding of contracts. Tendering is an important phase in procurement, and while also a wide-ranging term, it specifically refers to the bidding process, obtaining a price, and how a contractor is then appointed.

5.3 Procurement in construction

If procurement is the act of obtaining goods or services, in the construction industry, these goods or services can differ depending on the RIBA Stage of Work. For example, at Stage 0: Strategic Definition, a client may want to procure design advice, project management, planning and cost consultancy services. Further professional services may need to be procured at Stage 1: Preparation and Brief, so that the client has more information with which to make a decision. This can include sustainability assessments, risk analysis services and site surveying specialists. Further services, and in greater detail, progressing from feasibility to detailed technical information will then occur as a project progresses through Stage 2: Concept Design, Stage 3: Spatial Coordination and Stage 4: Technical Design.

During these RIBA work stages, depending upon the procurement route, one or more specialist subcontractor and/or a main contractor may also be procured to offer practical advice and design and management services. Also dependent upon the procurement route there may be an overlap between Stage 4 and Stage 5: Manufacturing and Construction. While services still may be procured, it is at these stages that the bulk of the construction materials will be purchased. Some may have been procured at earlier Stages if there was a high degree of certainty combined with long lead-in times or a strongly fluctuating market.

The 'iron triangle' is often associated with construction procurement, in that the goods and services are awarded against three criteria; time, cost and quality (the iron triangle). Historically, the construction industry has a reputation for focusing on the lowest cost, constantly overrunning in time and delivering a finished product that does not meet the desired quality. However, such perceptions are beginning to change as research has shown clients are now increasingly happy with the quality of finished projects, that are being delivered on time, and to the agreed budget (Watts et al., 2016).

Therefore, when procuring construction projects and services, clients are arguably looking for more in the way of additional benefits, such as increased ethical behaviour of those awarded contracts. This can take the form of agreements to follow approved codes of conduct or the inclusion of CSR and SV practices (as discussed in Chapter 4). However, it is not just the demands of clients that are increasing in their ethical focus, but also the behaviour of clients themselves as they procure tier one contractors and construction consultants during the RIBA Stages 0–4.

5.4 Ethical procurement

Evidencing ethics in procurement should in theory be relatively easy and straightforward to achieve. However, there are numerous examples of unethical procurement practices which can lead to entire procurement exercises being repeated, with great time and expense wasted for the party leading the procurement. For many, ethical procurement is focused on 'downstream' supply chain management and ensuring the supply chain is behaving ethically.

According to the Chartered Institute of Procurement and Supply (CIPS), establishing acceptable and consistent organisational behaviour is important to embedding ethical standards and ensuring they are followed, as is creating a culture where unethical practices are reported, and where managers have the skills and knowledge required to successfully understand and enact codes of conduct and legislative requirements (CIPS, 2015). CIPS also have Global Standards for Procurement and Supply with the five rights to procurement. These are cost, quality, time, quantity and place. However, there are also many ethical codes that are applicable to construction procurement. One code that can be applied to procurement behaviour is developed by the International Ethics Standards Coalition. With over 100 organisations under its umbrella, the coalition aims to have a global set of ethical standards for all professionals operating in the construction, property, land, infrastructure and related professions. These standards offer a consistent approach for all professionals, regardless of regional, economic or business differences (IESC, 2016). The latest standards were released in December 2016, and in the authors' opinions how they could apply to procurement practices within a construction contractor are discussed below. Although, it should be noted that the IESC do acknowledge that standards are an evolving process and will be continuously reviewed and updated as appropriate.

5.4.1 Accountability

Those who have ultimate responsibility for procurement activities should ensure that all rules and regulations are in place, and that all staff are trained and empowered to take responsibility for their own procurement behaviour. Directors should be held accountable to ensure rigorous and robust practices are in place, but individual commissioners, and those who complete the actual tasks involved in the procurement are accountable for their own actions.

5.4.2 Confidentiality

Procurement exercises almost always deal with confidential information. This can include everything from the names, addresses and details of those expressing an interest in submitting a tender, and those who have perhaps declined to submit, to pricing structures used by those tendering. This information needs to remain confidential, and cannot be shared among contractors tendering, nor can it be shared among colleagues without permission from the tenderer.

5.4.3 Conflict of interest

Any professional involved in procurement should consider and declare all conflicts of interest that may exist. For example, an ex-colleague, friend or family member may work for one of the contractors tendering, or you may have some connection with the company itself. Anything of this nature should be declared to management, and if the conflict can be removed, then it should be. If it cannot, then one of your current colleagues should assume responsibility for that element of the procurement. If that option is not feasible, then all decisions you make need to be fully transparent, with all decisions made double-checked, and all parties made aware of the conflict, with approval to continue sought.

5.4.4 Financial reporting

Having responsibility for the procurement of any work packages or services will result in the reporting of the final decisions made. The reporting of returned tenders needs to be accurate and truthful. This will have more emphasis if some tenders are missing information or have made mistakes in their submission, or if the comparison process is complex due to weightings for both financial and non-financial factors. Anyone with responsibility for procurement will need to ensure all received tenders are fully understood, recorded accurately, and any awards or proposals are made on the correct basis.

5.4.5 Integrity

Being honest, truthful and fair in all dealings, questioning instructions if you are not comfortable with them and speaking out against unethical practice evidence construction professionals' high integrity. Having integrity in procurement is of the upmost importance as it is a situation in which contracts worth potentially multi-millions of pounds are awarded. Acting with integrity also involves ensuring all advice offered is accurate and within your own competence.

5.4.6 Lawfulness

Ignorance of the legislation is not a valid excuse for not following legal requirements. It is a professional responsibility to ensure an awareness of all applicable and appropriate legislation, in the territory the procurement is taking place or will impact upon. This may prove difficult if, for instance, international regulations and legislation apply and it is the first time a construction professional has found themselves in such a circumstance. Nevertheless, all legislation must be understood and followed.

5.4.7 Reflection

Reflection is important in every task a construction professional will have to undertake. It is a skill that needs to be practised in order to be improved, and reflection on procurement is no different. A professional needs to consider what went well, and why, and what did not go so well, and why, during any procurement exercise. This is so lessons can be identified and learnt from, to prevent mistakes being repeated and any issues that arose during the procurement which were not identified and planned for beforehand can be addressed.

5.4.8 Standards of service

To offer a high standard of service professionals need to operate only within their competence with professionals ensuring they are suitably qualified to undertake the required services. This also extends to all staff involved in the procurement. It is not enough to simply delegate tasks to non-qualified members of staff, everyone involved needs to operate to a high degree of competence and ensure a professional and high-quality service is ultimately offered to clients on behalf of employers.

5.4.9 Transparency

Professionals need to ensure all practices undertaken are open and accessible for review. Information should be kept in a clear and secure manner, but care should be taken as you may not be able to share all documents with all parties, due to confidentiality issues. However, you must be transparent with a clear rationale for all decisions, and be able to clearly evidence this if required, to a party with the relevant right and authority. For example, the process by which decisions were made needs to be fully transparent so all parties have an equal and fair opportunity. This could take the form of an open and fully accountable procurement system, with clear guidance and instructions, that all parties are aware of from the outset. However, the individual pricing details of one tenderer are not to be shared with other competitors.

5.4.10 Trust

Embedding trust in all processes and instilling trust in clients, partners, colleagues and the wider supply chain is crucial. This is a standard that is earned through evidenced behaviour and practices, and, once built, is susceptible to loss if practices are not followed correctly. However, trust is key among all stakeholders to promote the reputation of a construction professional, their organisation, and the wider construction industry.

Following all of the above is important to evidence ethical procurement practices. Numerous different ethical standards and codes of conduct are available. If there ever was a claim of unfair contract award, evidence of following a code of conduct, values, or rules and regulations your organisation has in place will go a long way to reinforce your position and will prove that correct procedures have been followed. In addition to any internal codes you have followed specific to your company, an awareness of appropriate legislation is also a requirement to ensure all legislative requirements have been achieved.

5.5 Legislation governing procurement

There are many overlapping pieces of legislation which govern procurement. These cover public and private sector works, the anticipated duration of the projects, the expected overall costs, the nature of the works and the location where the works will take place. Since the United Kingdom (UK) has now left the European Union (EU), some pieces of EU procurement legislation will not apply in their current form in the future. However, during the transition periods the legislation will still apply, and if it is made into UK legislation during or after this period, then it will also still apply to UK procurement in future. There are also other examples of UK legislation that governs procurement behaviour. How ethical frameworks apply to the behaviour of professionals working within these procurement frameworks is discussed. Adherence to such legislation evidences the minimum ethical standards and behaviour expected of constriction organisations.

5.5.1 OJEU

The *Official Journal of the European Union* (*OJEU*) is the publication in which goods and services procured by the public sector (with values over set thresholds) must be advertised. It allows suppliers from any participating country to submit tenders for the advertised projects, with rules and regulations governing how the procurement exercises must be executed and how the tenders should be submitted and reviewed. Around 160,000 invitations to tender are published every week, with the production of the *OJEU* the responsibility of the Publications Office of the European Union (OJEU, 2020).

5.5.2 The 2014 EU Public Procurement Directive

This Directive applies to any public body procured works that exceed the financial thresholds set. If the anticipated costs of a project do not meet the thresholds set, the EU rules on transparency, non-discrimination, mutual recognition, equal treatment, and proportionality will still apply (Crown

Commercial Service, 2016). The Directive requires all *OJEU* tenders to be electronic and all procurement documents to be electronically available for all parties to review at the time the project notice is published.

The Directive provides for five different procurement procedures to be followed:

1. An open procedure – whereby all parties interesting in submitting a tender may do so.
2. A restricted procedure – only those parties who respond to the *OJEU* project advertisement are invited to submit a tender.
3. A competitive dialogue procedure – a selection is made from those who respond to the *OJEU* advert, with the public body client then entering into discussion with the interested parties to develop a solution which the chosen bidders will be invited to tender.
4. A competitive dialogue with negotiation – those who respond to the advertisement are asked to submit a tender. Negotiations may then take place to negotiate an improved offer.
5. An innovation partnership procedure – a selection is made of those who respond to the advert and negotiation then takes place for the submission of ideas for innovative working practices if there is no existing suitable product or service.

All procurement procedures are governed by the same rules, so that all suppliers tendering for the works are treated fairly and equally, regardless of country of origin. Any contract must also be awarded on the basis of the Most Economically Advantageous Tender (MEAT). This can, however, include the tender offering the best value for money, the best price/quality ratio and also take into consideration any applicable social and environmental criteria.

5.5.3 The Competition Act (1998)

With sections that reflect similar provisions within EU legislation, the Competition Act (1998) governs competitive practices within the UK. The aim of the Act is to prevent any business practice being undertaken that could have a negative impact on competition, Examples include large companies taking advantage of their size to prevent the growth of smaller competitors, those that have a dominant position in a market abusing that power, or even one company dropping their price to an unsustainably low level.

In the construction industry, an example of a practice that may fall foul of the Competition Act (1998) could be if you receive quotes from different contractors that are all similarly very high, and one quote that is lower than the others, but still higher than expected. This could be an example of price fixing, and is more prevalent in oligopolies, where only a small amount of competition exists, for instance, if there are only a handful of specialist contractors, or contractors of a certain size suitable to price for the works required. In such a situation, the contractors could agree among themselves who will submit the lowest tender, with all others agreeing to submit much higher tenders. This would allow one contractor to win the work at an inflated price, and in theory this 'favour' could then be repaid by the same situation happening in future, but with one of the other contractors submitting an inflated, but lower than the competition, tender price.

Although it is organisations that are involved in such practices, it is staff at these organisations who are ultimately responsible for such practices taking place. Therefore, it is construction industry professionals who need to be aware that any practices similar to those described above are illegal under the Competition Act (1998). Practices such as these are also unethical. Even if work is agreed to be 'shared out' among those in the oligopoly, the client is still losing out, and paying higher tender prices due to 'price-fixing' practices. Any similar behaviour from construction organisations is classed as a cartel, and there have been numerous examples evidenced over the past few decades which have damaged the reputation of the industry and left many clients suspicious of current construction industry practices.

5.5.4 The Bribery Act (2010)

With the primary aim of eliminating situations of bribery across the UK, the Bribery Act (2010) came into force on 1 July 2011. The Act is intended to reduce instances of bribery in a pragmatic way that does not increase pressure on smaller organisations to change their practices and procedures in a costly and time-consuming way.

According to the 2014 report, 'Fighting Corruption and Bribery in the Construction Industry' by the consultancy firm PricewaterhouseCoopers (PwC), construction is the industry that is the 'most affected by bribery and corruption'. The same report argues that

> the nature of the construction industry, where the procurement of goods and services and the selection of contractors and suppliers on large-scale projects may be decided or influenced by individuals within an organisation, provides a number of opportunities for corruption and bribery.
>
> (PwC, 2014)

It is, therefore, of the upmost importance during procurement that ethical integrity is always upheld by the construction professionals involved. At a basic level, this requires compliance with all relevant legislation. Therefore, in situations of potential bribery, it is important to be aware of the exact governance of the Bribery Act (2010).

While the Bribery Act (2010) covers many organisational behaviours and actions, in the context of procurement, it can be applied to many situations professionals face. For example, Section 1 governs the bribing of another person – not necessarily in the form of a financial gift. Construction professionals will need to take extra precaution to ensure that non-financial gifts are also not constituted as a bribe, for example, an alcoholic drink at Christmas or an invitation to a social event just before a procurement decision is made. Section 2 of the Bribery Act is the passive offence of accepting a bribe, such as the acceptance of a gift or experience event. Section 6 is the bribing of a foreign public official and Section 7 relates to an organisation failing to prevent bribery from occurring.

Discussion 5.1

Do you think it is acceptable for a project manager working for a main contractor to accept a bottle of wine from a project manager employed by a subcontractor on the last day of work before Christmas?

While the purpose of the Bribery Act is not to eliminate hospitality or restrict social and team building events and away days, it is to ensure these are done to serve a specific and intended purpose, which is not to gain additional work or favour in current and future business dealings. To help construction professionals navigate through any situations that may give rise to their ethics being questioned, the Bribery Act (2010) Guidance document (Ministry of Justice, 2010) outlines six principles that should be considered by professionals and organisations wishing to prevent bribery from occurring (Table 5.1).

The key for any professional involved with procurement is to evidence independence of thought and action. A professional needs to be aware of the

Table 5.1 Six ethical principles to prevent bribery

Principles	*Description*
1 Proportionate procedures	The action taken should be proportionate to the size and nature of the business and any risk faced
2 Top-level commitment	Higher management-level employees and owners are best placed to ensure a company follows the correct processes and has relevant procedures in place
3 Risk assessment	Any potential situations that may give rise to a bribery offence occurring should be suitably risk assessed
4 Due diligence	There should be an awareness of all people operating with or on behalf of a company
5 Communication (including training)	There should be clear channels of communication with regards to what the policies are on bribery, what kind of situations staff may face and how they should act, Suitable training for all staff should also be undertaken.
6 Monitoring and review	As businesses change, so do the risks they face, as well as the procedures they have in place. There should be an ongoing process of monitoring and review so that all practices and responses are in line with appropriate guidance.

relevant standards and legislation, and act in accordance with these. Actions undertaken must be done in a fair and just manner that professionals believe to be correct, and professionals should not be susceptible to amending behaviours due to pressure exerted from any other party.

5.5.5 The Public Services (Social Value) Act (2012)

Introduced as a Private Members Bill in 2010 by the then MP Chris White, and coming into force in January 2013, the Public Services (Social Value) Act (2012) (SVA) compels public bodies to consider broader social value when procuring goods and services over set threshold values. The Act is encouraged to be used in the procurement of projects, goods and services of all types and values, but must be considered in procurements valued over certain thresholds.

A requirement of the SVA is those commissioning any works to consider, at the pre-procurement stage, how improvements in the social, environmental and economic well-being of a relevant area can be achieved. This does pose several ethical issues that are perhaps new to consider for most professionals, such as considering what is meant by social value. This is a subjective definition, and so will mean different things to different people as discussed in Chapter 4. While this has many benefits, it also has many disadvantages. For example, if you are comparing returned tenders, and each tender has interpreted social value in a different way, you are then faced with having to ethically and fairly compare different social value activities. It is therefore of the upmost importance that clear guidance is provided to each interested party before they return their tender, so they know the weighting and scoring mechanisms associated with each of the procurement criteria. The type of social value expected should be considered in detail before tender information is issued to interested parties so you can avoid any potential ethical issues.

5.6 Social value in procurement and tendering

Although a relatively new concept in procurement and tendering for most, social value is already embedded in the procurement and tendering practices of many construction clients and contractors, most notably those who operate in the public sector and so are governed by the Public Services (Social Value) Act (2012). Traditionally construction procurement has revolved around the iron triangle of time, cost and quality, with quite often cost being the main driving force with clients' and contractors, with desire for 'cost effectiveness' surpassing all other criteria. However, leaders of some Local Authorities have confirmed that contractors are now expected to complete construction projects within the agreed timescales, for no more than the agreed budget and to the agreed quality – and so therefore the focus is now shifting as to how contractors can deliver such social value (Watts et al., 2016).

It is largely at the direction and discretion of the client as to how such social value should be judged as well as any requests and comparison mechanisms implemented in individual procurement and tendering scenarios. An increased focus on social value in procurement will inevitably lead to more questions being asked regarding the ethical practices employed by those managing the procurement, as there are two distinct aspects of ethical practice occurring simultaneously. Those that apply to procurement requirements, and those that apply to procurement practices,

5.6.1 Procurement requirements and ethics

This is where those with procurement responsibility are attempting to be more ethical in their procurement requirements, as they are arguably encouraging supply chain partners who are tendering to become engaged with wider issues, such as how they can provide and enhance social value. Therefore, with a focus on other criteria than simply time, cost and quality, those responsible for procuring are attempting to be more ethical, and also potentially encouraging the increased ethical practices of others. Simply stated, procurement can be used as a mechanism to increase the ethical practices of others.

5.6.2 Procurement practices and ethics

This is where the entire procurement process itself is conducted in an ethical manner. The professionals with responsibility for administering the procurement process will ensure all rules governing the procurement are effectively communicated to all parties involved. All timescales and approaches adopted in requesting and sharing information will also be adhered to, as well as the rules governing the comparison of tenders and calculation of final scores. Both transparency and confidentiality will be provided where appropriate.

Both ethically sound procurement requirements and practices will be required to ensure any procurement activities are completed to a high standard and that maximum social value is achieved. A fair and robust process that can be evidenced at each step of the journey will also be able to withstand scrutiny. If accusations of unfair decisions or unethical procedures are made, then there will be an increased focus on the decision-making processes that were followed to see if the decisions made were done in an ethically sound manner.

This is increasingly important, the more social value factors are adopted in procurement criteria, as there are no firm definitions or agreements as to what social value specifically relates to, with a degree of ambiguity actively encouraged to allow innovative and tailored social value activities to be adopted (Cabinet Office, 2015). However, it is vitally important in order to maintain ethical procurement practices that such ambiguity is then not passed down the supply chain during the procurement process. If each organisation tendering has to interpret and understand the social value requirements in their own way, this will inevitably lead to those with responsibility for the procurement having to compare elements of tender returns that are simply incomparable. Therefore, the discussion on what social value relates to, and what social value outcomes need to be achieved during the project or work package, must happen during the preparation of the procurement documents and before any information is made available for interested parties to use in their tender compilations. Clear advice and the parameters of social value will allow those organisations submitting tenders to engage with social value in a manner that can easily be communicated. Having social value activities that can easily be communicated and understood will enable those managing the procurement to easily compare different tenders, to reach decisions on awarding contracts easier and in a transparent and logical manner, having followed an approach that is robust enough to withstand scrutiny. It will also allow those tasked with managing the successful delivery of the project or work package to clearly measure the compliance and success of the agreed social value activities.

Discussion 5.2

You work for a consultancy advising a public sector client on the procurement of a main contractor to construct a primary school on a design and build basis. How would you compare the following social value activities from three different main contractors' returned tender submissions?

1 Contractor A: The creation of 100 apprenticeships and weekly litter picking of a local park for the duration of the project.
2 Contractor B: The creation of 50 apprenticeships and a commitment to 95 per cent local spend.
3 Contractor C: A commitment to train 20 unemployed individuals in a trade of their choice, ensuring each one secures a CSCS card, a full-time job, and enrolment on a college course.

5.7 Social value communication and measurement

The communication and measurement of social value are of the upmost importance to the successful delivery of social value. Without clear communication, all stakeholders may interpret social value intentions differently and so may not understand exactly what is intended. Without clear measurement, it will be difficult, if not impossible, for it to be accurately understood if the intended social value has been achieved.

Progress has been made over the last decade in both the theoretical literature and pragmatic industry approaches to the communication and measurement of social value. However, a range of approaches exist that both complement and contrast with one another, which although they have served to further the debate and progress in some areas, they have also served to restrict progress in others (Watts, Dainty and Fernie, 2019). It is fair to say that the concept of social value is growing in importance with requirements increasing in procurement criteria. Therefore, participation with social value as a concept is no longer optional for those construction organisations wishing to work somewhere in the public sector supply chain. Many private sector clients are also increasingly using social value criteria in the procurement of their own projects and services, driven in part by wider stakeholder awareness and demand, and also the internal values and intentions of the private sector clients themselves. Research has also shown that the private sector often follows the lead of the public sector (Snider et al., 2013).

While this textbook does not focus on the communication or measurement practices of social value, it is important to understand the increasing importance social value will have in construction industry procurement, as social value is intrinsically linked to ethical practice. Therefore, ethics will play an increasing role in the criteria used during procurement, and not just the procedures and practices followed.

5.8 The benefits of social value in procurement

It is important to briefly note the many benefits of including social value in procurement. These benefits can be experienced by all stakeholders involved, immediately and in the medium and long term. For example, if an organisation adopts social value in their procurement, they will be increasing the amount of social value impacts experienced as a result of their organisational actions. Similarly, for those unsuccessful in their tender submissions, they will ultimately have to engage more with social value activities in order to increase their likelihood of procurement success. For construction organisations that are successful in tenders, they will then have to deliver the promised level of social value.

In addition to the intended recipients of any social value activity, i.e. the apprentices hired, the business owners benefitting from local spend, and the community enjoying a renovated and tidied local landmark or garden, social value activities also benefit the organisations involved in their delivery. Studies have shown that those organisations embracing social value and the related concepts experience a more positive reputation and general business image among stakeholders (Du, Bhattacharya and Sen, 2010). These organisations also benefit during the recruitment of staff as they are more likely to have a wider range of applicants due to an increased organisational appeal (ibid.). Staff are also more likely to stay with an organisation as an increased focus on social value is linked to higher staff retention levels (Brammer, Millington and Rayton, 2007). Ultimately, and perhaps unsurprisingly as social value is linked to procurement, studies have also shown those organisations who engage with social value and the wider CSR concept are more likely to increase their financial performance due to all the mediating variables that are positively impacted (Saeidi et al., 2014).

It can, therefore, be stated that an organisation who believes social value engagement and activity are part of behaving ethically, will be rewarded for their ethical endeavours, both by the positive impacts they encourage or administer in society and by the increased business benefits they experience. However, in order to achieve the latter, any organisational ethical behaviour, and compliance with existing requirements or social expectations, will need to be evidenced, both during procurement and beyond.

5.9 Evidencing ethical compliance

Applying the rules as dictated by any procurement body, piece of legislation or formal directive will assist in the evidencing of an ethical approach undertaken in any construction procurement activity. However, some ethical standards are easier to evidence than others. Compliance with set rules, for example, can easily be recorded, with all decisions transparently made and evidence kept, thus adherence to the rules set can then be clearly illustrated if ever questioned.

This is perhaps easier to evidence when procurement is all about simple set criteria. For example, if works are to be awarded on the basis of lowest cost, proposed timescales, or a mixture of the two. If clear and transparent weighting is given to the procurement criteria, and a shared and open scoring mechanism is discussed among all those taking part in the tender, then the process of calculating the successful tender is relatively straightforward. Those construction professionals with responsibility for the procurement will then be able to transparently apply the scoring mechanisms to determine the successful tender, with all unsuccessful parties being notified in a clear and transparent manner.

However, illustrating ethical compliance may be more difficult when it is complex construction projects or services that are being procured and when the procurement criteria are less well defined, or wider than the iron triangle of cost, time and quality, such as when the Social Value Act is used in public procurement. In such circumstances, clear rules of tendering will need to be communicated to all parties involved in returning tenders, and closely followed by professionals involved in the management and delivery of the procurement. Agreed metrics (where possible) and/or model tender submissions should always be identified by the party with responsibility for the procurement. Discharging responsibility for social value requirements to the supply chain who are tendering, without a clear and consistent focus and articulated vision, will inevitably lead to a messy and difficult-to-follow procurement process. This could ultimately leave an organisation open to complaints and legal action, if one unsuccessful tenderer feels the tender process was not clear. For example, if one party claims inconsistent information was received, or if their submission scored less than a competitor, and how the scores were to be awarded was not clearly described beforehand. This is a simple trap to fall into for those with responsibility for procurement, especially when they may be unsure of exactly what they want to achieve and so use ambiguous social value criterion.

Those with procurement responsibility should ensure the procurement route used best suits their needs and the information they have at the time of starting the procurement process. The selection of an incorrect procurement route will hinder the clarity of information tenderers require and ultimately make it more difficult to clearly evidence an ethically sound process has been followed.

5.10 Evidencing social value in different procurement routes

In the construction industry, a procurement route refers to the method of procurement undertaken. The strategy adopted to determine the method will need to take full consideration of the client's requests, resources and requirements. The suitability of procurement routes is a key consideration in the overall success or failure of a project, as often incorrect client advice at an early stage (or the lack of any suitable advice sought) can reveal numerous expected and unexpected problems later on in the construction process. High standards of ethical practice are required throughout any procurement undertaken, whether it be of consultants and professional advice at RIBA Stage 0, the procurement of a main contractor and/or specialist subcontractors at RIBA Stages 3/4, or the procurement by the main contractor of subcontractors at Stage 5.

There are many different procurement options available, all of which require careful ethical consideration. There are many commonalities among the ethical considerations of different procurement options, and also ethical considerations that are specific to the circumstances each procurement option presents. Several procurement options and their relationship with ethics are considered below.

5.10.1 Open tendering

Open tendering is when a client (used in this sense to describe any party procuring the goods and/or services of another) posts an invitation to tender. Any party that is interested in the works may then return their tender by the date required. Dependent upon the size and scope of the works being tendered, the process may be more complex and require a PQQ (pre-qualification questionnaire) stage. This is a stage that separates all those parties who express an interest in returning a tender into those the client deems suited to return a tender, and those not suited. PQQs can include questions on the locality of the tenderer, their size, recent experience, turnover and information on the tenderer's company policies and practices.

While open tendering does allow for maximum competition and in theory allows any company the chance to express interest, it is also a time-intensive and costly process for all involved. This includes the companies who have to review and assess each PQQ and each tender received, and also for all the companies who are completing the PQQ and submitting tenders, as the majority will be unsuccessful. While this is an inevitable part of the construction industry, it is perhaps unethical to have such a process available for a wide selection of submissions, when the majority will not be suitable for your company's needs. This sometimes happens when companies want to project an image of openness and state that they agree to work with any tenderer suitable. However, in practice, the company may only advertise this to project an image and will have no intention of working with the vast majority of those companies to submit tenders.

Discussion 5.3

Is it ethically correct to allow numerous companies to go to the time and expense of completing PQQs and submitting tenders if you have no intention of placing an order with them?

5.10.2 Selective tendering

Selective tendering is similar to open tendering in that an invitation to tender is published by the party acting as the client. However, in this procurement method, a 'short list' of potential tenderers is drawn up who are then sent the invitation tender. This often occurs when there is only a selection of potentially applicable parties available based on their experience, location, turnover, etc. This initial list then reduces, further based on those who are able to submit a tender in the time frame provided and then again when a selection is finally made, and the contract awarded.

Selective tendering also takes the form of approved supply chains, with organisations often required to complete some form of pre-qualification exercise to be on an 'approved list' for the purposes of selective tendering, Alternatively, some main contractors for example, have an approved supply chain based on those subcontractors with whom they have successfully worked previously. While this does allow a certain degree of confidence and can help foster a collaborative environment between main contractors and subcontractors, it can also prove very difficult for new suppliers to be considered for tenders.

Such an approach to tendering also raises several ethical issues. First, there needs to be clear information available for how new subcontractors apply to get put on a selective tenders list, and transparency over how and why contractors are selected for the invitation to tender. A lack of clarity over the contractor requirements that obfuscates the processes involved can lead to questions being raised over an organisation's ethical practices. Even if practices that make it difficult for new suppliers to be added to selective tendering lists are unintentionally used, this still evidences a poor ethical oversight on behalf of the organisation involved.

Discussion 5.4

If a main contractor only issued invitations to tender to the same subcontractors without a robust and transparent qualification process in place that allows new subcontractors to join, is the organisation committing unethical practice?

5.10.3 Framework tendering

A framework tender is where a construction organisation will complete the procurement requirements for a client and will then be entered onto a list of approved companies for one or more clients for the set amount of time the framework exists. The clients will then go directly to one of the approved companies for future work. The rules of any framework can differ from one framework to the next. Being on the approved list may offer a guarantee of future work, or it may mean that further procurement and tendering are required on a project-by-project basis.

The benefits for construction companies can include limiting the amount of resources needed for procurement and tendering. If projects have to be bid for individually, it can be time-consuming and repetitive, whereas if you only have to bid once to be on a framework, you can employ your workforce more efficiently. It can also offer a guarantee of a certain workload for the duration of the framework which can then result in economies of scale for purchases and the incentive for construction companies to offer more cost-effective solutions, embrace partnership approaches, and invest in longer-term social value.

Framework agreements can be implemented by a single client or can be developed and managed by external companies on behalf of one or more clients. One example of such a framework agreement is that of the Scape Group. The Scape Group offers frameworks on behalf of predominantly public sector clients. In 2018, Scape procured a civil engineering framework worth up to £2.1bn over a four-year period, which was awarded to Balfour Beatty. This framework covered all those public sector clients who signed up across England, Wales, Scotland and Northern Ireland and allowed the clients to complete any civil engineering works in a shorter duration as the pre-construction procurement activities were all undertaken in a single exercise. The benefits of such frameworks across the whole of Scape activities include an advertised half a million-pounds saving for their collaborative procurement approach over the traditional alternative, with social value benefits of 27,000 apprentice hours, 7,000 work experience hours and diverting over 840,000 tonnes of waste from landfill (Scape Group, 2020).

Frameworks are similar in approach to selective tendering but with more formal rules in place and can lead to large and more frequent projects. However, the ethical scrutiny required will be the same. All processes will need to be transparent, with awarding criteria and weightings clearly identified for all parties. A third-party framework provider that serves to connect clients and contractors together can also help confirm robust and independent ethical practices have been followed.

5.10.4 Negotiated tendering

In practice, a negotiated tender can occur at any time. This could be at the end of an open or selective tendering process, with all the tenderers informed that upon appointment a process of negotiation will be entered into to finalise all details and prices. Negotiated tender is also a standalone procurement route whereby a single contractor will be approached by a client with the intention of negotiating the terms of the project before entering into a contract for the works. This procurement option may not be available in many instances, such as those where a competitive process needs to be clearly evidenced, for example, this applies in many cases where an element of public funds are being used. However, where negotiated tenders can occur, they do provide benefits such as encouraging a collaborative approach, and the focus of resources as both parties are confident a full contract agreement will be entered into.

Although by the same reasoning, ethical issues can also increase as, first, the selection of a party to negotiate with will be scrutinised, especially if problems are revealed at a later date, such as cost and time overruns – people may ask why this procurement route was selected. Second, during the negotiation itself, both parties could adopt a more adversarial position due to the lack of competition and mistrust instead of allowing a more collaborative ethos to develop. If this procurement option is elected, all parties will need to ensure ethical guidelines are followed so that decisions made can withstand scrutiny should questions arise at a later date. Ethically, if there were no other contractors that met the client requirements, and a negotiated procurement process suited the clients' brief, then a negotiated procurement approach would be ethically valid. During the negotiation stage, however, ethical frameworks would have to be adopted to guide the process to ensure a transparent and easy-to-follow decision-making and agreement-reaching process had occurred.

5.11 Payment practices in construction

Payment terms in the construction industry relate to the number of days a payment will be made by one company to another (i.e., client to consultant, client to main contractor, main contractor to subcontractor). While the issue of payment may appear relatively straightforward and easy to follow in theory, in practice, it is an area of great ethical dilemma and one that needs to be navigated carefully.

It was Lord Denning who once famously stated that cash flow is the life blood of the building industry (Riches, 2016). In this statement Lord Denning was referring to the importance of cash flow to sustain companies operating in the industry and the importance of making prompt payments throughout the supply chain so all parties can access monies owed and continue to operate.

It is often late payments, or simply no payments being made at all, that are the main reason for the majority of insolvencies of companies operating within the construction industry. Insolvency can be described as an organisation being unable to satisfy outstanding debts when they are due or having more liabilities than assets on their balance sheet (Jones, 2019). While insolvency itself can be a temporary process and is described as a 'state of being' (a state which many companies in the construction industry constantly operate in), being insolvent does not itself mean a company enters liquidation or administration. However, problems are often compounded by the fact that due to the highly competitive nature of the industry, many companies operate on very low profit margins in order to be successful at winning tenders which are awarded on a lowest cost basis.

Discussion 5.5

When awarding a contract, is it ethical to award based on purely the lowest cost submitted, even if you know that the lowest price is 'too' low to achieve the quality required

Payment practices in the construction industry are notoriously poor, in that the date payments are made is often later than contractually agreed, with the contractually agreed dates often months after the work has taken place. Ethically, it is not correct to prolong payments to supply chain members longer than is required, however, by the very nature of the construction industry, payments cannot be made immediately and there is always going to be a lag between works carried out and payment made. Holding onto monies for longer than required or using such monies to fund other activities is, however, an unethical practice and should be avoided by construction organisations. The slow processing of payments in the construction industry and calls for it to improve due to its unethical nature and serious financial ramifications for the supply chain are nothing new, with many highly public examples evident over recent decades, from Lord Denning calling for better payment practices in the construction industry in the 1970s to the publication of 'Constructing the Team' by Sir Michael Latham in 1994 (often called the Latham Report) which stated, 'payments should be made promptly and on time' (Latham, 1994, p. 20). This was then echoed in 2016 by Mark Farmer with the publication of 'The Farmer Review of the UK Construction Labour Model' (often referred to by its subtitle; 'Modernise or Die'), which stated the discussions held as part of the review reported the 'payment practices within industry as poor, with a tendency to rely on extended payment terms' (Farmer, 2016, p. 25).

In an attempt to address these long-standing payment criticisms, there has been an increasing move by construction clients in requesting that construction companies sign up to the Prompt Payment Code (PPC). The PPC was launched in 2008 (although more recent moves have been made to get the construction industry involved post the publication of the Farmer Review). The Chartered Institute of Credit Management (CICM) helped launch the PPC and is a professional body for credit management professionals. They are the largest professional body of their kind in the world. The aim of the PPC is to improve the culture of prompt payments, with signatories to the PPC agreeing to pay 95 per cent of invoices within 60 days while working towards paying within 30 days (Prompt Payment Code, 2020). Signatories also have to agree to the following conditions:

1 Pay all suppliers on time in accordance with the agreed timescales.
2 Provide clear guidance to all suppliers regarding the terms of payments, with clear communications if payments are to be late and guidance on dispute resolution.
3 Encourage participation of the PPC throughout their own supply chain.

Any organisation which does not follow these practices is removed from the list of those who participate and abide by the PPC. Reinstatement is possible after removal but requires a remediation plan to be approved by the PPC's compliance board prior to reinstatement. Those who have been removed can be publicly named and many construction clients want any company they work with to be a signatory to the PPC, and so removal from the PPC due to frequent late payments can ultimately lead to a lack of future procurement opportunities (CICM, 2020).

The 2014 EU Public Sector Procurement Directive is implemented in the UK by the Public Contract Regulations (2015) and took effect on 1 September 2016. These regulations place a duty on public bodies to pay undisputed invoices to supply chain members within 30 days. It is hoped the practices of the public sector can serve as an example of best practice for the private sector. The Regulations actually state that Central Government should pay 80 per cent of undisputed invoices within five days, Local Authorities should aim to pay invoices within 10 days and suppliers can claim statutory interest on unpaid invoices after 30 days from receipt of invoice (Crown Commercial Service, 2015). It is hoped that such prompt payments can then be filtered through the supply chain with all suppliers paid on similar payment dates in an attempt to keep cash moving.

Discussion 5.6

Is it unethical to extend supply chain payment terms far beyond those what your own organisation has in place with your client?

There is also further legislation that aims to encourage fair supply chain payment terms in the form of the Small Business, Enterprise and Employment Act (2015). Part 1, Section 3 of this Act states that companies should publish information regarding '(a) the company's payment practices and policies relating to relevant contracts of a prescribed description' and '(b) the company's performance by reference to those practices and policies' (Small Business, Enterprise and Employment Act, 2015). This legislation impacts all companies in some way, and a record is kept of construction company's performance against the requirements of the Act by BuildUK (2020). The table, published and updated every six months by BuildUK, ranks construction companies on the following three criteria:

1 The percentage of invoices not paid within agreed terms.
2 The average number of days taken to pay invoices.
3 The percentage of invoices paid within 60 days.

With the legislative requirements, publicly available databases and best practice codes of conduct being used by clients, it is hoped that extended and unethical payment practices will be a thing of the past. Indeed, many construction companies pay all invoices on time and advertise this fact, which is good for individual company ethical behaviour, but arguably negative in that it reflects the wider practices of the industry if prompt payment of the supply chain is still a distinguishing factor between companies. Nevertheless, such steps are promising in that payment practices may begin to change to reflect more ethical company values. However, there are ramifications for the industry if ethical payment practices, and broader ethical procurement behaviours generally, are not widely adopted.

5.12 The ramifications of unethical procurement practice

For construction professionals, ethical procurement should be common practice. Although as outlined throughout this chapter, situations can occur where unethical decisions are made during procurement. Sometimes such decisions can be arrived at with the best of intentions at the start of the process, but through a variety of reasons and events a procurement exercise can end with a decision made unethically. However, sometimes more sinister circumstances can occur in which intentional unethical decisions can be made during procurement. In both circumstances the outcome is essentially the same, a procurement decision is made unethically and so the ramifications will ultimately be the same. Such ramifications can include the following scenarios.

5.12.1 The de-skilling of industry professionals

If organisations are successful in their procurement attempts due to unethical practices, such as engaging in bribery, the winning of projects will not depend upon the skills and experience of staff. Therefore, an organisation engaged in bribery may be less likely to develop the skill sets of their existing staff. This will also then have knock-on effects for those organisations who are unsuccessful in their procurement attempts due to a competitor's bribery. The unsuccessful organisations will have less income and so less profit to reinvest in their own staff. Ultimately, bribery on any scale serves to restrict the development of construction professionals, but the more prevalent it is, the wider the ramifications will be felt throughout the industry.

5.12.2 A failure to invest in research and innovation

Much like the failure to invest in the development of staff, companies that gain a competitive edge through unethical practices also are less likely to invest in research and innovation. This lack of investment will also be prevalent among those companies who are not successful in their submitted tenders due to the unethical practices of competitors. While those who are initially unsuccessful may invest in further innovative products and methods in the hope of achieving success in procurement, if this success is not forthcoming, due to continued unethical competitor practices, then ultimately there will be no resources for companies to invest in developing innovative approaches.

5.12.3 Reputational damage

Reputational damage is a two-part problem. The first is to the company involved in unethical practices. Inevitably, unethical practices are always revealed. In some instances, there may be a long period of time between the unethical practice occurring and it finding its way into common knowledge, or sometimes the process may be shorter. Either way the outcome is ultimately the same, the reputation of those involved will be tarnished indefinitely as they are associated with unethical practices. The process of 'damage mitigation' required will depend on the severity of the ethical transgression and whether it is an isolated incident or reflective of deeper and more widespread systemic practices. The second part of the problem from a reputational point of view is that it reflects badly on the entire industry and therefore encourages clients to treat the industry with an element of distrust and acts as a barrier to the recruitment of new operatives and professionals.

5.12.4 Removal of chartered professional status

For those who are chartered professionals, unethical practice can have very real and immediate ramifications as they will be removed from professional memberships. This can be both a professional embarrassment in the fact that such a removal of membership is made public and will be associated with their name in the industry for any future employer to find. It may also therefore restrict future employment opportunities and lead to the termination of current employment.

5.12.5 Insolvency, administration and liquidation

Ultimately, if employees engage in unethical practices, either on behalf of their organisation or if acting independently, and as a result an organisation finds it difficult to win work or recruit new staff due to their negative reputation, they may eventually end up becoming insolvent and then being forced to enter a process of administration or liquidation. This could have positive benefits in that such organisations are no longer operating in the construction industry and so opportunities for ethical organisations should increase. However, rarely is an entire organisation unethical, but usually only a small group of employees.

These employees could be at any hierarchal level, but if an organisation were to cease trading due to the unethical behaviour of a few employees, then it would be unfair and unfortunate for all the hard-working and ethically behaving employees who suffer as a result.

5.13 Summary

This chapter has illustrated the evolution in procurement criteria from a process largely focused upon time, cost and quality, to the inclusion of wider procurement criteria, such as CSR and SV. The increased focus on ethical decision making this requires is discussed as are the long-standing ethical dilemmas and scenarios encountered during construction procurement. The role ethics plays in all manner of procurement decisions is identified, as are the methods by which ethical compliance can be evidenced. Finally, the ramifications of unethical decisions during procurement are summarised in the hope that construction professionals understand the implications of any decisions made during procurement and have greater insight into the ethical considerations needed at every stage.

References

Brammer, S., Millington, A. and Rayton, B. (2007). The contribution of corporate social responsibility to organizational commitment. *The International Journal of Human Resource Management*, 18 (10):.1701–1719.

BuildUK (2020). Construction sector payment performance. Available at: https://builduk.org/priorities/improving-business-performance/duty-to-report/table/

Cabinet Office (2015). Social Value Act Review. Available at: https://assets.publishing.service.gov.uk/government/uploads/system/uploads/attachment_data/file/403748/Social_Value_Act_review_report_150212.pdf

Chartered Institute of Credit Management (2020). CICM-originated Prompt Payment Code moves to Small Business Commissioner's Office. Available at: www.cicm.com/cicm-originated-prompt-payment-code-moves-small-business-commissioners-office/

Chartered Institute of Procurement and Supply (2015). Global Standard for Procurement and Supply. Available at: www.cips.org/Documents/Global_Standard/Global_Standard.pdf

Crown Commercial Service (2015). Public Contracts Regulations 2015: Statutory guidance for contracting authorities and suppliers on paying undisputed invoices in 30 days down the supply chain. Available at: https://assets.publishing.service.gov.uk/government/uploads/system/uploads/attachment_data/file/555393/Revisedstatutoryguidance26Sept.docx.pdf

Crown Commercial Service (2016). A brief guide to the 2014 EU Public Procurement Directives. Available at: https://assets.publishing.service.gov.uk/government/uploads/system/uploads/attachment_data/file/560261/Brief_Guide_to_the_2014_Directives_Oct_16.pdf

Du, S., Bhattacharya, C. and Sen, S. (2010). Reaping relational rewards from corporate social responsibility: The role of competitive positioning. *International Journal of Research in Marketing*, 24(3): 224–241.

Farmer, M (2016). The Farmer Review of the UK Construction Labour Model: Modernise or die, time to decide the industry's future. Available at: www.constructionleadershipcouncil.co.uk/wp-content/uploads/2016/10/Farmer-Review.pdf

IESC (International Ethics Standards Coalition) (2016). The standards. Available at: https://ies-coalition.org/standards/.

Jones, P. (2019). How insolvency affects companies in the construction sector. *The Gazette Official Public Record*. Available at: www.thegazette.co.uk/insolvency/content/103418

Latham, M. (1994). Constructing the team: Joint Review of Procurement and Contractual Arrangements in the United Kingdom Construction Industry. Final Report. July 1994. Available at: https://constructingexcellence.org.uk/wp-content/uploads/2014/10/Constructing-the-team-The-Latham-Report.pdf

Ministry of Justice (2010). The Bribery Act 2010: Guidance. Available at: www.justice.gov.uk/downloads/legislation/bribery-act-2010-guidance.pdf

Official Journal of the European Union (2020). What is the OJEU? Available at: www.ojeu.eu/Default.aspx

Prompt Payment Code (2020). What is the PPC? Available at: www.smallbusinesscommissioner.gov.uk/ppc/about-us/

PwC (2014). Fighting corruption and bribery in the construction industry. Available at: www.pwc.com/gx/en/economic-crime-survey/assets/economic-crime-survey-2014-construction.pdf

RIBA(Royal Institute of British Architects) (2020). RIBA Plan of Work 2020. Available at: www.architecture.com/-/media/GatherContent/Test-resources-page/Additional-Documents/2020RIBAPlanofWorktemplatepdf.pdf

Riches, J. (2016). In search of the Holy Grail, Part Two. Adjudication Society. Available at: www.adjudication.org/resources/articles/search-holy-grail-part-two

RICS (Royal Institution of Chartered Surveyors) (2014). RICS professional guidance, UK: Tendering strategies. Available at: www.rics.org/globalassets/rics-website/media/upholding-professional-standards/sector-standards/construction/black-book/tendering-strategies-1st-edition-rics.pdf

Saeidi, S., Sofian, S., Saeidi, P., Saeidi, S.P. and Saeidi, S. (2014). How does corporate social responsibility contribute to firm financial performance? The mediating role of competitive advantage, reputation, and customer satisfaction. *Journal of Business Research*, 68: 341–350.

Scape Group (2020). Measured performance: Simple procurement options that create real value. Available at: www.scapegroup.co.uk/scape-in-numbers

Snider, K., Halpern, B., Rendon, R. and Kidalov, M. (2013). Corporate social responsibility and public procurement: How supplying government affects managerial orientations. *Journal of Purchasing and Supply Management*, 19: 63–72.

Watts, G., Dainty, A.R.J. and Fernie, S. (2016). *The influence of public sector procurement practice in shaping construction CSR*. Paper presented at: RICS Annual Construction, Building and Real Estate Research Conference (COBRA 2016), Toronto, 19–22 September.

Watts, G., Dainty, A.R.J. and Fernie, S. (2019). Measuring social value in construction. In C. Gorse (ed.) *Proceedings of the 35th Annual ARCOM conference*, 2–4 September 2019, Leeds, UK, Association of Researchers in Construction Management.

6 Professional bodies and ethics

6.1 Introduction

This chapter will articulate why there needs to be ethical principles for those professionals working in the construction industry and will explain the importance of such principles. It will also highlight the challenges for ethical teaching, given that there are no universal standards around the world and standards vary between different professional institutions on what constitutes good ethical practices.

The chapter will explain codes of conduct in regulating professional ethics, what these comprise and how they are applied to improve practices and behaviours. In this regard, it will discuss the benefits of having codes of conduct and how they vary between different professional bodies and institutions. Penalties and sanctions for non-adherence and breaches of professional codes of conduct will be covered, alongside the implications for public trust and confidence of violations of ethical regulations. Alongside this, the importance of maintaining public trust will be discussed, given that the construction industry traditionally has been regarded as poorly performing in the past with low levels of client satisfaction. Conversely, the reputational benefits of strict compliance with institutional codes of conduct will be articulated and why these are important for maintaining professional standards.

Embedding codes of conduct and ethical behaviour and standards into organisational culture by professional bodies will be examined alongside the factors which are influential in promoting best practices. In this regard, strategies and models will be identified and how they are maintained within organisations will be discussed. Finally, several professional bodies will be compared in their approaches to professional ethics, standards and behaviours, and the differences between their codes of conduct explained.

6.2 Ethical principles for construction professionals

Owing to increasing concerns in many high-profile cases which demonstrate dishonesty and corruption, it is important for construction professionals to commit to and encourage project teams to comply with sustainable ethical principles. Codes of ethics which have been introduced provide an indicator that organisations and institutions take ethical principles seriously as they outline expectations for all personnel with regard to ethical behaviour and intolerance of unethical practices (CMI, 2013).

Relationships between construction clients and the professional consultants and contractors they appoint rely on professional ethics and trust, especially since fee agreements cannot accurately specify all the financial contract contingencies for possible additional services (Walker, 2009). The main reason why the public relies on members of professional bodies relates to them giving advice and practising in an ethical manner (RICS, 2010). Accordingly, the RICS has developed eight ethical principles to assist their members in maintaining professionalism and these relate to honesty, openness, transparency, accountability, objectivity, setting a good example, acting within one's own limitations and having the courage to make a stand. In order to maintain the integrity of the profession, members are expected to make a full commitment to these values.

Arguably the main deficiencies of codes of ethics have arisen from the notion that there are no universal standards and accordingly they vary between countries and different sectors in the building industry. Boundaries and barriers created by fragmentation and differentiation within the construction sector have possibly deterred the emergence of any common frameworks of professional ethics in the past (Walker, 2009). This is an area that demands more attention through multinational dialogue across all areas of the construction sector.

6.3 Codes of conduct to regulate professional ethics

Members of professional bodies are bound by codes of ethics, sometimes referred to as ethical principles, to address the issue of non-ethical behaviour and to attempt to provide a context of governance (Liu et al., 2004). Members of such institutions are usually bound by codes of conduct in the way they practise, and the institutions reserve the right to take action against members who breach the rules and regulations laid down. A professional code of conduct can be described as the minimum level of behaviour expected of an individual member of a profession. Normally such behaviour relates to professional practice and compliance with institutional rules and regulations and the aim is to preserve the reputation and good name of a profession through self-governance.

Almost every profession has its codes of conduct to provide a framework for good ethical choices (Abdul Rahman et al., 2007). In the United Kingdom and indeed worldwide, construction industry professionals have qualifying and professional institutions and bodies that represent each discipline, such as the Royal Institute of British Architects (RIBA), the Chartered Institute of Architectural Technologists (CIAT), the Chartered Management Institute (CMI), the Royal Institution of Chartered Surveyors (RICS), the Chartered Institute of Building (CIOB) and many more. These institutions have strict charters, professional codes of conduct and ethics, rules and regulations relating to professional standards that individual members are required to follow. These may differ in some detail from one institute to another, but the majority of them do require members to act professionally and use and apply common ethical and moral principles in conducting business. Professional institutions who introduce and administer the codes of conduct for their members will normally enforce sanctions and impose disciplinary action on their members for non-compliance. If these individuals fall short of the required standards, they could be subject to a disciplinary committee, who in some cases can expel them. For serious cases, possibly involving illegal activity, e.g., fraud, the institution could report the perpetrators to the relevant authorities for further action.

Most institutes and organisations have separated codes of ethics from codes of conducts, and some, like the CMI, define their codes of ethics as 'a statement of the core values of an organisation and of the principles which guide the conduct and behaviour of the organisation and its employees in all their business activities' (CMI, 2010). One of the reasons that a code of professional conduct is applied is to promote good conduct and best practice (RIBA, 2005).

It is normal for professions such as the Royal Institution of Chartered Surveyors to enforce mandatory continuous professional development (CPD) on their members to ensure they keep abreast of skills, education and regulations. These are linked to the professional institutions' ethical commitments to keep their members attuned to new developments in the sector, especially in their specialist areas of competence and experience. Some of the criticisms of codes of conduct in the past have been voiced at the constraints and limits of the codes and their ability to manage behaviours, especially in the construction industry.

Technological and scientific advances in recent decades have led to continued requirements for the review, update and introduction of new codes of conduct by institutions and organisations. Such measures are required in order to respond to changes to aspects of human activities and environmental changes, such as global warming and other environmental issues. The construction industry arguably has more interest in environmental issues than other sectors, possibly due to the building processes which have traditionally been heavy on carbon generation. In addition, in recent years there has been an increased focus on the construction life cycle costs of building, and so energy efficiency and use of renewable technologies have become more important.

It is important for construction professionals to be aware that adopting the aforementioned codes of conduct, gains and maintains respectability and integrity for project teams, and their respective professional institutions and organisations (Vee and Skitmore, 2003). Stewart (1995), however, was sceptical of such institutional codes of conduct and explained that these are merely guidelines for professionals to interpret as they wish and believes they do not promote values, ethics and morality accordingly. Clearly corrupt behaviour is subject to more than just policing by professional institutions and in some cases can be deemed a criminal offence. This being the case, there is a strong argument to suggest that the law, coupled with institutional sanctions, may present an acceptable level of deterrence in that professionals will think twice before committing unethical behaviour if the consequences are considered grave enough. Liu et al. (2004) argued, however, that transgression will still occur if detection of breaches is considered unlikely or if the disciplinary measures imposed for breaches are regarded as insufficient or too lenient. Conversely, there is an argument that ethical codes of conduct should not be regarded negatively as a framework for punishing breaches but positively in assisting professionals in recognising their own moral parameters (Henry, 1995). Accordingly, this raises the question of the role of codes of conduct and whether their purpose should be more closely linked to promoting compliance. This could involve motivating professionals to behave in an ethical manner as opposed to them be seen as frameworks to impose punishments for potential breaches.

6.4 Maintaining high standards of professional conduct and competence

When considering codes of conduct, one needs to determine not only the particulars relating to as particular code but also the context of what the code is intended for and its ability to comply with it. Introducing codes of conduct, codes of ethics, rules and regulations by professional institutions is a strong message and indicates to members what is expected of them in terms of behaviour. It also sends a strong signal to other stakeholders that unethical practices will not be tolerated (Fewings, 2009).

Codes of conduct developed by professional institutions are designed to provide a strict code of adherence to ethical values, in which the interests of communities and clients take priority over the self-interest of individuals belonging to those professions (Haralambos and Heald, 1982). Therefore, the primary reason for having codes of conduct is to ensure and embed governance and regulation of professional ethics. This is an important focus for the book when considering that relationships between clients and their professional contributors have relied in the past on a significant level of trust and professional ethics (Walker, 2009). Furthermore, the heart of best practice in construction management is the maintenance of high standards of professional conduct and competence, underpinned by the principles of honesty and integrity (CMI, 2010). This view was reinforced by Vee and Skitmore (2003) and they articulated the importance of codes of conduct as the tool to enable such standards. They argued that construction professionals should have the fundamental commitment to professional conscience and professional competence. Professional competence in this context can be defined as the capability to perform the duties of one's profession generally, or to perform a particular professional task with skill of an acceptable quality (ibid.). Furthermore, professional competence is predicated on having the broad skills, attitude and knowledge to work in a specialised profession or area. Disciplinary knowledge and the application of concepts, processes and skills are required as the measure of professional competence in any particular field. Professional competency is one of the five fundamental principles of professional ethics along with integrity, confidentiality, professional behaviour and objectivity. Notwithstanding this, each of the professional institutions has their own lists of technical and professional competencies, which are closely aligned to their institutional rules and regulations.

Another important role for codes of conduct, according to Lere and Gaumnitz (2003) is to influence decision making. Notwithstanding this assertion, the problem has been to establish those codes of ethics which are most effective at promoting ethical decision making. This involves determining those codes of ethics that impact decision making with the effect that they change the values and beliefs of individuals. Furthermore, according to Inuwa et al. (2015), codes of conduct are to ensure clients' interests are properly cared for, while wider public interest is also registered and protected. The RICS have articulated that as clients' expectations have changed, business practices need to change also to respond to these. This could mean that ethical issues need to feature high on the list of organisational strategies and business drivers. Codes of ethics established by professional institutions, such as the RICS, can serve as 'checks and balances' for individual members to try to curb unethical or immoral behaviours. This could assist in reducing undesirable practices and behaviours on a global level.

6.5 Governance and enforcement of professional ethics through codes of conduct

In practice, as mentioned above, professional ethics are policed by a national or international body to ensure a minimum standard of practice for organisations to strictly adhere to. Such bodies are most frequently professional institutions that seek to ensure that the behaviour of construction professionals is controlled and monitored by a strict code of ethics, which they create and maintain (RICS, 2001). These codes of ethics, frequently referred to as codes of conduct, are primarily to ensure that integrity and ethics are maintained at all times. Institutions such as the RICS and RIBA have royal charters, which strictly set down rules and regulations relating to professional standards, morals and ethics which all members must comply with. Members of professional bodies are bound by codes of conduct to address the issue of non-ethical behaviour and to attempt to provide a context of governance (Liu et al., 2004). Members of such institutions are then bound by such codes of conduct in the way they practise, and the institutions reserve the right to take action against members who breach rules and regulations laid down.

Professional ethics through the aforementioned codes of conduct set down acceptable norms and thresholds relating to standards of practice. This creates ethical codes, codes of conduct and sets of rules by which member individuals or organisations are regulated. Furthermore, measures are taken to give more transparency relating to the investigation of what could be deemed behaviour or actions of an unethical nature. This adds robustness and due diligence to the process and could have the effect of boosting public confidence. Codes of conduct can sometimes be synonymous with rules to deal with certain issues and particular situations. Examples could include members of a professional institution maintaining indemnity insurance and processes for reporting the malpractice of fellow members. The institutional rules and regulations more often than not cover potential conflicts of interest at an individual or corporate level. This would normally apply where a person or organisation has obligations to more than one party on a commission or contract and from either a private or commercial standpoint there is a clash of interests. This has to be considered from the position of trust that lies with that particular individual or organisation and measures should be taken to ensure that others are not disadvantaged by a lack of impartiality.

There can be deterrents to avoid or reduce unethical behaviour and ensure adherence to codes of conduct and these can be intrinsic to individuals. However, professional institutions or organisations can also impose extrinsic deterrents to such behaviours through enforcement provisions, and these may be made compulsory or voluntary. Individuals will still decide for themselves whether to comply with such codes and so they cannot be forced to comply with them. However, professional institutions or organisations can impose strict penalties for codes which are violated. How individuals view such penalties will affect the extent to which they comply. For instance, if individuals regard the penalties as having negative consequences for them, then they will be more likely to comply and the codes in such cases become compulsory for them. Examples of such penalties could be instant dismissal from an organisation or expulsion from a professional institution. Alternatively, if the decision makers regard the penalties as less serious for them, then even the heaviest penalties will not make compliance compulsory.

6.6 Misconduct and the reputation of professions

Construction projects have traditionally had a reputation for poor performance, coupled with low levels of client satisfaction (Latham, 1994). Poon (2003) argued that there is a specific lack of research into the ethical issues of construction management that could result in less than satisfactory project outcomes. This is especially the case as professionals rely on both knowledge and ethical conduct so that the general public will have confidence in them. For a professional institution to gain public confidence in this regard, maintaining ethical practices are very important, as they have a direct influence on the quality of those services provided to clients and the public image and perception.

Where misconduct and unprofessional behaviour have been identified in the past, this has had negative implications on the way the construction industry has been viewed. In some cases, this has led to public concern and government attention and action. Conversely, where there is deemed to be a high level of ethical practices adopted and adhered to by a profession, then this would normally imply high performance and low client dissatisfaction levels. When specifically considering the surveying profession, one needs to consider those services not specifically related to construction projects but the whole diverse range of property services across the spectrum over the life cycle of buildings. In the past, the RICS has focused on trying to improve the reputation of the institution in this regard and distinguish it from other services, such as estate agency. This was in a concerted effort to improve trust by the public in light of some cases of professional misconduct. Notwithstanding this position, Poon (ibid.) argued that the focus for training surveyors has traditionally been linked to providing core skills and 'hands on' knowledge required to perform their services. Such priorities are not always focused on improving the quality of those services provided and ethical considerations in some cases have been regarded as a sub-strand to working practices rather than a foundation those services are built upon. Perhaps this explains why in the past there have been calls for reforms to training and education programmes for surveyors and how these may assist in improving quality levels (Ashworth, 1994, as cited in Poon, 2003).

6.7 Embedding ethical codes, behaviours and standards into organisational culture

Liu et al. (2004) argued that, in the past, in the construction industry there has been a focus on culturally related attributes of project teams and a general commitment to quality. Notwithstanding this premise, they concurred that there has been evidence to suggest that ethical behaviours and standards are responsible for successful project outcomes.

Professional ethics can be reinforced through professional standards of professional bodies and codes of practice to gain respectability and integrity. However, these codes and standards need to be embodied in management practices and roles and responsibilities assigned to ensure compliance across ethics programmes. As such, these codes are insufficient in themselves to ensure ethical conduct and accordingly require to be complemented with organisational responsibilities and employer training programmes. Positions such as ethics officers could be instrumental in 'turning the tide' in bringing ethics and ethical values to the forefront of organisational values. Such interventions and measures could shape how organisations are structured and how they function and would bring reputational benefits for them. A survey carried out by Vee and Skitmore (2003) of construction professionals in Australia, found that most companies (90 per cent) had subscribed to a professional code of ethics, although only 45 per cent had 'gone the extra mile' and committed to ethical codes of conduct. Most of the organisations and individuals who contributed to the survey agreed that ethical practice is considered to be an important goal for their respective organisations. Furthermore, most of them agreed with the view that business ethics should be governed and driven forward by personal ethics.

Professional codes of practice and conduct are committed to by members of institutions as a prerequisite for being a member of that institution. According to Stewart (1995), they should, however, not be used to teach ethics, values or morality. They simply are there to lay down rules of conduct as guidelines for action. Enforcement of the codes can be an important factor, and if there are few or no consequences, then it is unlikely that individuals will always act in the way that a particular institution intends. However, when consequences are grave for a particular breach, people will think twice before infringing ethical codes of conduct. A similar phenomenon is apparent when the risk of detection of breaches is low. If individuals believe they can breach ethical codes of conduct and get away with it, with little chance of being found out, then it is more likely that they will commit unethical practices. In this case it can create an environment of opportunism where ethical behaviours are heavily influenced by the organisation. In this way ethical practices and behaviours can be delineated by the climate and culture of the organisations and the boundaries of socially acceptable norms in which they operate. Ethical codes do not in themselves always solve moral dilemmas or reduce instances of illegal practices. However, they work best where they are supported by mechanisms and structures that ensure review, adjudication, communication, oversight, and enforcement.

Liu et al. (2004) created an organisational ethics model, particularly focused on the perspectives of surveyors and using data obtained from the Hong Kong Institute of Surveyors (HKIS). It analysed several factors under the categories of ethical climate, culture and codes to assess which of the factors were more influential and important in promoting ethical standards and behaviours than others. These are listed in Figure 6.1.

It was concluded from the research by Vee and Skitmore (2003) that, from the perspective of surveyors, the dominant factor in both the public and

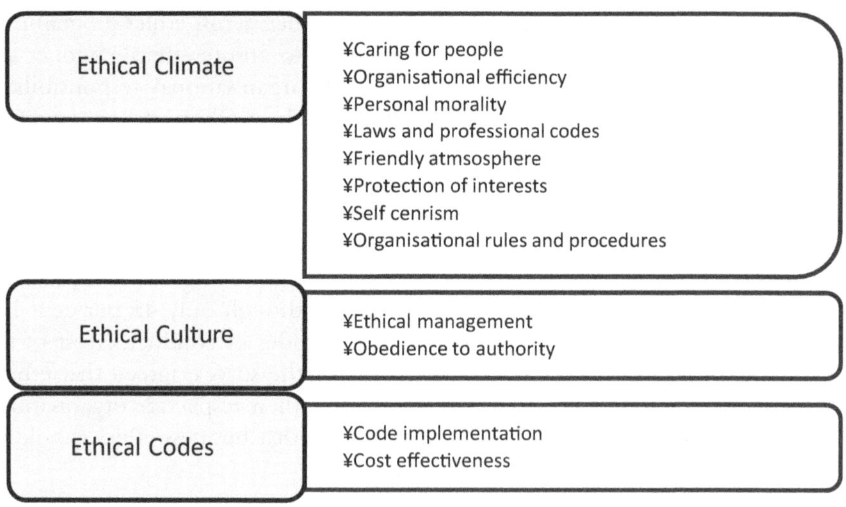

Figure 6.1 Model of ethical climate, culture and codes
Source: Adapted from Liu et al. (2004).

private sectors was the law and the professional codes Surveyors, as professional practitioners, conformed more to professional institutional standards which act as behavioural norms for their practice. Furthermore, the study showed that ethical codes are more effectively implemented in the public sector, where financial and behavioural governance play a very important role in organisational management. These codes are deemed to be less clear in the private sector where their use in some cases could be more related to enhancement of reputation and image. For codes to be effective, it is crucial for them to be communicated down throughout the whole organisation.

6.8 The Royal Institute of British Architects (RIBA) Codes of Conduct

The Royal Institute of British Architects (RIBA) have three principles for their members to strictly adhere to and these are honesty and integrity, competency and relationship. These are built upon the institutional values of concern for others, concern for the environment, honesty, integrity and competency (RIBA, 2005). These principles are contained in their Code of Conduct in Table 6.1 and are supported by clear rules and regulations around competition, advertising, insurance, complaints and dispute resolution. In addition, the RIBA provide guidance notes covering aspects such as maintaining integrity and confidentiality. The Institute has Disciplinary Procedure Regulations that it administers, specifically related to breaches of ethical principles and regulations. It imposes disciplinary measures for contraventions according to the code, and professional misconduct can be dealt with through sanction powers and a RIBA hearing panel.

6.9 The Chartered Institute of Building (CIOB) Codes of Conduct

The Chartered Institute of Building (CIOB) sets out its codes of professional conduct based on the institution's and members' duties to clients, the supply chain and employees. Furthermore, the CIOB recommends and advocates strongly that services must be performed to a high-quality standard and should achieve value for money for clients. They also reinforce the importance of professionalism and duties owed to the general public to enhance the reputation of the construction industry.

The CIOB Rules and Regulations of Professional Competence and Conduct (CIOB, 2019) consists of 14 rules governing what members should demonstrate, undertake and fulfil in discharging their duties with complete fidelity and probity. They also include four regulations related to the use of distinguishing letters, logo, advisory service and advertising. Figure 6.2 contains some rules related to complying with their conduct for CIOB members in this regard which are predicated on the common premise of competencies, trust, respect, professionalism and honesty.

Table 6.1 The RIBA Code of Conduct

1.	*Principle 1 Honesty and Integrity*
1.1	The Royal Institute expects its Members to act with impartiality, responsibility and truthfulness at all times in their professional and business activities.
1.2	Members should not allow themselves to be improperly influenced either by their own, or others', self-interest.
1.3	Members should not be party to any statement which they know to be untrue, misleading, unfair to others or contrary to their own professional knowledge.
1.4	Members should avoid conflicts of interest. If a conflict arises, they should declare it to those parties affected and either remove its cause or withdraw from that situation.
1.5	Members should respect confidentiality and privacy of others.
1.6	Members should not offer or take bribes in connection with their professional work.
2.	*Principle 2 Competency*
2.1	Members are expected to apply high standards of skill, knowledge and care in all their work. They must apply their informed and impartial judgement in reaching any decisions, which may require members to balance differing and sometimes opposing demands (for example, the stakeholders' interests with the community's and project's capital costs with its overall performance).
2.2	Members should realistically appraise their ability to undertake and achieve any proposed work. They should also make their clients aware of the likelihood of achieving the client's requirements and aspirations. If members feel they are unable to comply, they should not quote for, or accept, the work.
2.3	Members should ensure that their terms of appointment, the scope of their work and the essential project requirements are clear and recorded in writing. They should explain to their clients the implication of any conditions of engagement and how their fees are to be calculated and charged. Members should maintain appropriate records throughout their engagement.
2.4	Members should keep their clients informed of the progress of a project and of the key decisions made on behalf of the client.
2.5	Members are expected to use their best endeavours to meet the client's agreed time, cost and quality requirements for the project.
3.	*Principle 3 Relationships*
3.1	Members should respect the beliefs and opinions of other people, recognise social diversity and treat everyone fairly. They should also have a proper concern and due regard for the effect that their work may have on its users and the local community.
3.2	Members should be aware of the environmental impact of their work.
3.3	Members are expected to comply with good employment practice and the RIBA Employment Policy, in their capacity as an employer or an employee.
3.4	Where members are engaged in any form of competition to win work or awards, they should act fairly and honestly with potential clients and competitors. Any competition process in which they are participating must be known to be reasonable, transparent and impartial. If members find this not to be the case, they should endeavour to rectify the competition process or withdraw.
3.5	Members are expected to have in place (or have access to) effective procedures for dealing promptly and appropriately with disputes and complaints.

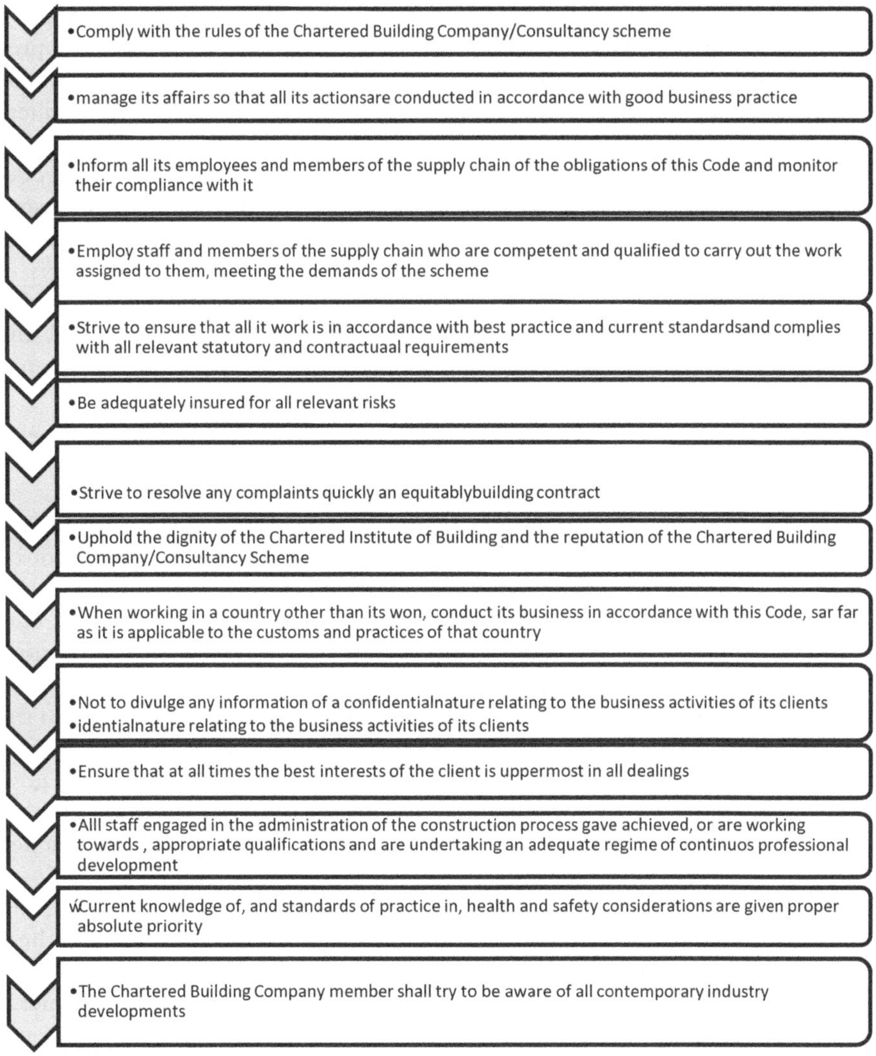

Figure 6.2 Rules relating to the CIOB codes of conduct

6.10 Summary

For a professional institution to gain public confidence, maintaining ethical practices is very important, as they have a direct influence on the quality of those services provided to clients and the public image and perception. Codes of ethics can be described as a statement of the core values of organisations and of the principles which guide the conduct and behaviour of organisations and their employees in all their business activities. Members of professional bodies are bound by these codes of ethics, sometimes referred to as ethical principles, to address the issue of non-ethical behaviour and to attempt to provide a context of governance. These institutions have strict charters, professional codes of conduct and ethics, rules and regulations relating to professional standards that individual members are required to follow and adhere to and the institutions reserve the right to take action against members who breach the rules and regulations laid down. Accordingly, introducing codes of conduct, codes of ethics, rules and regulations by professional institutions sends out a strong message and indicates to members what is expected of them and this can serve as 'checks and balances' for individual members to try to curb unethical or immoral behaviour. They can also send a strong signal to other stakeholders that unethical practices will not be tolerated, and this could assist in reducing the undesirable practices and behaviours on a global level.

In considering governance and regulation, professional ethics are policed by a national or international body to ensure a minimum standard of practice for organisations to strictly adhere to. Professional institutions or organisations can impose strict penalties for codes which are violated to deter unethical practices and behaviours. In this sense, however, ethical codes of conduct should not be regarded negatively as a framework for punishing breaches but positively in assisting professionals in recognising their own moral parameters. Clearly corrupt behaviour is subject to more than just policing by professional institutions and in some cases can be a criminal offence.

Where misconduct and unprofessional behaviours have been identified in the past, this has had negative implications on the way the construction industry has been viewed and, in some cases, this has led to public concern and government attention. To address this, there has been a concerted effort to improve the general public's confidence and trust after cases of professional misconduct. Professional ethics, reinforced through professional standards of professional bodies and codes of practice, have played an important part in regaining this confidence and trust alongside respectability and integrity. However, these codes and standards need to be embodied in management practices and roles and responsibilities assigned to ensure compliance across ethics programmes.

Codes of conduct are sometimes anchored in the Seven Nolan Principles and embedded as core requirements for organisational financial regulations and these are selflessness, integrity, objectivity, accountability, openness, honesty and leadership. Codes of ethics should reflect the practices and cultures which construction clients want to encourage for their respective organisations and project teams. The Code of Ethics Checklist devised by the Chartered Management Institute provides clear guidance for employees on what is expected of them in terms of ethical behaviour and practices. This sends a clear signal to other parties including customers and suppliers that unethical practices are not acceptable. To support their Codes of Conduct and Professional and Ethical Standards, the RICS have quite helpfully created a decision tree for proceeding or not proceeding with certain courses of action, and for decision making to maintain such standards.

References

Abdul Rahman, H., Karim, S., Danuri, M., Berawi, M. and Wen, Y. (2007). Does professional ethics affect construction quality? Available at: www.sciencedirect.com (accessed 24 November 2019).

CIOB (Chartered Institute of Building) (2019). Rules and regulations of professional competence and conduct. Available at: www.ciob.org (accessed 23 May 2020).

CMI (Chartered Management Institute) (2010). Code of professional management practice. Available at: www.managers.org.uk/code/view-code-conduct (accessed 24 November 2019).

CMI (Chartered Management Institute) (2013). *Codes of Ethics Checklist*. London: Chartered Management Institute.

Fewings, P. (2009). *Ethics for the Built Environment*. London: Routledge.

Haralambos, M. and Heald, R.M. (1982). *Sociology: Themes and Perspective*. Slough: University Tutorial Press Limited.

Henry, C. (1995). Introduction to professional ethics. *Professional Ethics and Organizational Change*, 13.

Inuwa, I.I., Usman, N.D. and Dantong, J.S.D. (2015). The effects of unethical professional practice on construction projects performance in Nigeria. *African Journal of Applied Research (AJAR)*, 1(1): 72–88.

Latham, M. (1994). *Constructing the Team*. London: The Stationery Office.

Lere, J.C. and Gaumnitz, B.R. (2003). The impact of codes of ethics on decision making: Some insights from information economics. *Journal of Business Ethics*, 48(4): 365–379.

Liu, A.M.M., Fellows, R. and Nag, J. (2004). Surveyors' perspectives on ethics in organisational culture. *Engineering, Construction and Architectural Management*, 11(6): 438–449.

Poon, J. (2003). Professional ethics for surveyors and construction project performance: What we need to know. In *Proceedings of the RICS Foundation Construction and Building Research Conference (COBRA), 1st September to 2ndSeptember 2003*. London: RICS Publications.

RIBA (Royal Institute of British Architects) (2005). *Code of Professional Conduct*. London: RIBA.

RICS (Royal Institution of Chartered Surveyors) (2000). *Guidance Notes on Professional Ethics*. London: RICS Professional Ethics Working Party.

RICS (Royal Institution of Chartered Surveyors) (2001). *Professional Regulations and Consumer Protection Department*. London: RICS.

RICS (Royal Institution of Chartered Surveyors) (2010). *Maintaining Professional and Ethical Standards*. London: RICS.

Stewart, S. (1995). The ethics of values and the value of ethics: Should we be studying values in Hong Kong? In S. Stewart and G. Donleavy (eds), *Whose Business Values?*, Hong Kong: Hong Kong University Press, pp. 1–18.

Vee, C. and Skitmore, M. (2003). Professional ethics in the construction industry. *Engineering, Construction and Architectural Management*, 10(2): 117–127.

Walker, A. (2009). *Project Management in Construction*. Oxford: Blackwell.

7 Pre-design and design stage ethical dilemmas

7.1 Introduction

This chapter is concerned with the application of ethical frameworks in practical construction industry-based scenarios as they relate to both the pre-design and design phases of construction projects. As ethics at this stage of the project often relate to financial decision making, the focus has been divided into three sections: Section 7.4 looks at financial project performance from the consultant's perspective. Section 7.5 explores the contractor's position, as would be applied in a procurement scenario whereby the contractor is involved in the early stages of the project. Section 7.6 explores more general ethical concerns. Finally, two real-world project case studies are explored to demonstrate the complexity of determining the boundaries of legal and ethical practice and the challenge of addressing unethical practice from general failings in project governance.

7.2 RIBA Stages 0–4

The Royal Institute of British Architects (RIBA) is the leading professional body for architects, established in 1837. In 2020, the latest RIBA Plan of Work (RIBA, 2020) was launched, comprising of eight work stages. Stages 0–4 deal with the processes of pre-design and design. These stages are: 0: Strategic definition; 1: Preparation and briefing; 2: Concept design; 3: Spatial coordination; and 4: Technical design. It is during these stages when the project scope is estimated, business cases established, and the design of the project developed. However, it is important to remember the RIBA Plan of Work is simply a framework through which the RIBA envisages and recommends construction decisions to be taken and this will evolve as the demands of procurement change. This chapter therefore focuses on the personal, organisational and professional ethics that could potentially emerge during these phases of a typical project, assuming a traditional or design and build procurement route is adopted, as these have been identified as dominant in UK practice (NBS, 2018).

7.3 Professional ethics

The following are a series of real-life examples which pose ethical dilemmas that you may face during your professional career. You may consider some to be small ethical dilemmas where the course of action you should take is clear, others are more substantial and complex. In most cases all possible courses of action have far-reaching consequences, both indirect and direct, immediate and in the future, that are not ideal, regardless of the choices you make. While so far this textbook has explored all manner of ethical frameworks from a somewhat theoretical perspective, this chapter takes these frameworks and applies them to a set of scenarios to help navigate the ethical dilemmas you may face or be asked to advise upon throughout your professional career.

Predominantly this chapter will focus on the ethical issues encountered by construction industry professionals during the execution of their daily responsibilities. For each scenario presented below, ethical standards and frameworks will be applied as worked examples that will guide best practice behaviour in a variety of ethical scenarios. Twelve questions will be posed and the answers given, on ethical decisions.

7.4 Theme 1 The financial management of projects at pre-design and design stages by the client's professional team

This theme explores several scenarios associated with project costing within the design phase of projects. Given the importance and ramifications of project costing, this section ends with a real-life case study showing the complexity of project costing and the difficulty of deciphering genuine error from unethical practice.

7.4.1 Scenario 1: Political pressure to resolve cost estimate

You are working as a quantity surveyor and have responsibility for developing the initial business case for a new sports centre. The sports centre represents a major development in a local regeneration area and replaces a very popular facility that had to be demolished due to extensive corrosion of the steel frame a couple of years ago. The new development has the support of external stakeholders, including the local community, who are desperate to have access to a local pool, after a new facility was constructed in a neighbouring town. Despite the desire to construct the pool, budgets are tight, and you are worried the funds available within the capital budget will not be adequate to meet the full costs of the scheme.

ⓘ *Question 1*

Working for the local authority you can see the level of public support for the new sports facility is growing and the pressure on local politicians to deliver the project is increasing. You are also aware that the local elections are scheduled for next May and the political group in control of the local authority are facing a decreasing majority and may struggle to resecure control of the council, so every vote matters. Given the desire to maintain local support, the leader of the council has put a lot of pressure on senior council officials to deliver a feasibility estimate which both fits within the current capital budget structure and ultimately adds support to the business case, allowing the project to proceed. Would you be happy to follow this direction and develop an estimate that fits the funding available and allows the project to proceed?

Answer 1

You may, rightly, feel under a lot of pressure to give the politicians the answer they want – that the local authority has the finances available to support this project. It is clear the senior member of the organisation is keen to keep the support of the elected members and to be seen to support the whim of their elected bosses and deliver this politically important project. Equally, you may well feel the sports centre is unaffordable and the estimate you develop will not fit into the tight financial envelope and that trying to make this appear to be affordable would be unacceptable and the project would end up costing significantly more.

From an ethical perspective, is this correct? At this point, I am sure you are of the opinion this is unacceptable, but let us consider both the RICS Global Professional and Ethical Standards (RICS, 2020) and the Rules and Regulations of Professional Competence and Conduct (CIOB, 1993). It is clear from a review of both standards that a number of breaches of professional ethics are associated with this scenario. The first of the five RICS ethical standards requires members to 'act with integrity', this is also a requirement for the CIOB, as rule 3 requires members to 'act with integrity so as to uphold and enhance the dignity, standing and reputation of the institute' (ibid.). Can we describe a construction industry professional who purposefully under-estimates the cost of constructing a project to allow that same project to proceed to design to retain the support of the local community for elected officials as acting with integrity? Regrettably, the literature (Flyberg et al, 2004; Cantarelli et al., 2010) reveals the answer to this question is quite often yes, with political interference and the resulting deliberate cost under-estimation identified as key triggers for post-contract cost over-run. While this is the case, this does not in any way excuse such conduct, the RICS guidance in terms of acting with integrity expressly advises that chartered surveyors should not allow 'the undue influence of others override their professional or business judgements and obligations'. It must be recognised at this point, that the political pressure applied in this scenario is difficult and going against this is not an easy decision, but it is equally clear that the RICS would expect you to resist this undue influence and provide an accurate and realistic cost estimate for this project regardless of the political fall-out. Equally, the RICS standards also stipulate that the surveyor has a duty to 'act consistently in the public interest when it comes to making decisions or providing advice'. Would it be in the public interest to allow an under-funded project to proceed in the knowledge that this represents a serious drain on public funds which would likely result in other equally important projects having to be cancelled to address the under-funding issue when the project reaches site?

? *Question 2*

Your senior QS has now been instructed that the project must be included in this year's capital budget, which is not fully committed, as a couple of less urgent projects have moved back to the next financial year. However, this still does not leave enough money, in your opinion, to complete the project. The budget currently suggests £10m is available, but you suspect the sports centre project, which has to include a water flume, will cost closer to £13m. Despite your initial reservations, your boss has issued you with a management instruction to proceed and develop an initial budget estimate (Order of Cost estimate) for the project that fits within the budget and will be acceptable when the business case is forwarded for approval. How would you proceed?

Answer 2

This is a very difficult position, and one you may well find yourself in at some point in your career. It is clear the senior management team are trying to keep a positive relationship with the elected members of the council and also that your line manager is happy to follow instructions given from the senior management team for whatever reason (we are not going to elaborate on this). So, you have a difficult choice to make at this point, often ethical decisions are not easy and do present in a challenging way, such as in this example.

Having looked at the issues this scenario raises, we can now turn to the ethical codes from our two professional bodies, starting with the RICS. In this scenario, a number of RICS Global Professional and Ethical Standards are relevant to this scenario, both from the perspective of the surveyor but also in terms of the conduct of the senior surveyor. First, as already discussed, the RICS ethical framework obliges us to 'act with integrity' (RICS, 2020) while the CIOB Rules and Regulations of Professional Competence and Conduct mandate their members to 'at all times act with integrity so as to uphold and enhance the dignity, standing and reputation of the Institute' (CIOB, 1993). In this situation, would the surveyor be acting with integrity if they misrepresented the anticipated cost of the scheme to fit within the client's budget? The simple answer is, no, they wouldn't. They would, however, be acting with integrity if they were to suggest the scheme as proposed would be too expensive and would not fit within the available budget, but if the project scope were reduced, this would then allow the scheme to progress.

The next ethical standard to be considered is 'act in a way that promotes trust in the profession' (RICS, 2020), a similar requirement informs those who hold chartership with the CIOB, in that they are expected 'in fulfilling their professional responsibilities and the duties which they undertake, to have full regard to the public interest' (CIOB, 1993). Therefore, if the surveyor were to proceed with the instruction and deliberately under-estimate the cost of the project, would this promote trust in the profession? Trust extends to all stakeholders in the project, not just the trust of the management team within the organisation who would consider your actions positively as an indicator of compliance with their instructions or indeed the elected members of the council, who would have been given the political answer they wanted, but the wider stakeholders. At some point, this project would go to tender and would the financial pressure of the budget be transferred to the contractor and thus some level of inappropriate value management be implemented, such as that implemented at Grenfell and highlighted in the Hackitt Review, where value engineering was seen to be less about value and more about cutting cost and quality to fit the budget (Hackitt, 2018)? Equally, how would the deliberate under-estimation of the likely cost of the project be perceived by the general public? It would hardly engender their trust in the profession.

Additional ethical duties emerging from the CIOB Rules and Regulations of Professional Competence and Conduct include the ethical duty to 'discharge their duties with complete fidelity and probity' (Regulation 5) and more specifically to ensure, when providing an advisory service, that the advice given is 'fair and unbiased' (Regulation 5.5) (CIOB, 1993). Clearly, if the surveyor chose to continue with the deliberate under-estimation of the work on this project, such a decision would be at odds with the requirements of both these duties of professionalism and ethical responsibility.

Discussion point 7.1

Having considered the ethics of the surveyor tasked with producing the feasibility costings for this scheme, do you feel the senior surveyor's conduct is ethical? In developing your answer, use the RICS Global Professional and Ethical Standards and the CIOB Rules and Regulations of Professional Competence and Conduct to inform your response.

In response to Discussion point 7.1, the same ethical obligations would apply to the senior surveyor, assuming they held chartered membership, but even if they didn't, the same ethical approach to professionalism would be expected by their peers, employers and members of the public. In this situation, the senior surveyor would be expected to do the following:

- Act with integrity (CIOB, 1993; RICS, 2020). In the same way as the surveyor would be expected not to deliberately misrepresent the anticipated cost of the scheme to fit within the client's budget, the senior surveyor would be expected to prevent this approach from being adopted.

- Promote trust in the profession (CIOB, 1993; RICS, 2020). Once again, if the senior surveyor encouraged or reported a deliberately incorrect estimate for the project to ease the political situation, this would hardly promote trust in the profession or the organisation's surveyors from a business ethics perspective.

Given the surveyor's seniority and position of power over the surveyor, there are also additional ethical considerations associated with such conduct:

- Take responsibility (RICS, 2020). The senior surveyor needs to take responsibility for the actions of junior colleagues especially when they are acting on their direct instructions. If there is a clear error rather than deliberate attempt to under-estimate the project, the surveyor should nevertheless take responsibility for this error as they have a professional duty to supervise the work of a junior colleague and check this is correct before it is issued to a client.
- Treat others with respect. Does the instruction to a junior colleague to deliberately manipulate and under-report the true cost of a project for political reasons really constitute treating others with respect? They should put themselves in the difficult position of trying to refuse to complete an unethical task and ask themselves, was that really an appropriate instruction?

7.4.2 Scenario 2: Under-staffing leading to error

You are working for a consultancy practice; at the moment the practice has taken on a lot more work than it can realistically cope with, due to an unexpected increase in repeat business and the commercial fear of turning away repeat clients. The situation has also been compounded by two members of staff resigning and another two taking long-term sickness absences. As a result, you have found yourself under pressure to prepare a number of financial reports for clients, develop a several cost plans for ongoing projects and to complete the tender documents for a final project. As a solution, you ask an assistant and trainee quantity surveyor to help out.

? *Question 3*

Is this approach correct?

Answer 3

At the moment, there is absolutely no ethical breach. It is important to allow those new to the profession to develop skills and competence, it is expected within the RICS APC framework that novice surveyors will gain practical experience. The key to this will be the levels of supervision and control put in place to ensure the client receives the best service possible.

⑦ *Question 4*

One of the team has produced the cost report, was too rushed to check it and with a tight deadline you issue the report and the client calls with concerns about the likely cost that seriously exceeds the cost limit you determined. What do you do?

Answer 4

As established, the approach of letting a junior colleague complete the task was, in itself, not unethical. However, as the more experienced surveyor, you also have the obligation to provide supervision and check the work is correct. This situation opens up to ethical concerns. The first is the duty to the client, we must not forget as construction professionals we have an ethical obligation to the client to 'Always provide a high standard of service' (RICS, 2020), which the RICS advises includes the importance of always ensuring your client receives 'the best possible advice, support or performance of the terms of engagement you have agreed'. In this situation, it is clear this has not happened, and this should not have happened.

However, now this has happened, the next ethical obligation is to 'take responsibility' (ibid.) for the error. Ultimately, it is the experienced surveyor's error in that they should have supervised and checked the accuracy and appropriateness of the advice issued. However, now the issue has emerged, you should advise your client of the error, outline your organisation's formal complaints process, if the client feels the need to formally register their concerns with the service received, and be accountable for the error. This includes doing something to address this error. The surveyor certainly should not turn around and blame their junior colleague for this issue.

The second part of this scenario is the ethical duty owed to colleagues and especially those colleagues who are less experienced or junior in position within the organisation. Was the surveyor right to ask a junior colleague to undertake the work? As outlined earlier, with the correct controls and safeguards in place, experiential learning is a fundamental part of professional training. However, in this example, those controls and safeguards had been overlooked, or not considered, so this was more a transfer of excess work. Was that ethical? The short answer is no. Within professional ethics, it is expected professionals will 'act with integrity' (ibid.). Was this allocation of work treating the junior colleague with respect? Within the RICS ethical standards guidance, it is stated that you should not 'take advantage of a colleague', which is arguably what has happened here.

It could be argued that the junior colleague also has an ethical obligation to 'act within their own competence' (ibid.), but it would have been very difficult for them to have determined the boundaries of that competence, especially as someone who is actively learning the role of a quantity surveyor. As such, they would have trusted their senior colleague to have given them work they felt was in their competence to achieve or at least ensured the learning process had been completed with feedback on their work and adjustment, to ensure it was correct before it was issued to the client. Furthermore, that senior colleague also has an ethical obligation to 'act in a way that promotes trust in the profession' and, as such, needed to be aware of the possible impacts of their actions on their colleague.

Ultimately in this context, the surveyor's decision to pass the task to a junior colleague, without any form of control or safeguards, whereby the work was checked and any adjustments actioned before it was issued to the client, has raised a number of ethical concerns.

7.4.3 The case of the National Paediatric Hospital, Dublin

So far, the ethics of this situation have been rather clear-cut, however, in the realities of practice they are often somewhat more blurred. For example, if we consider the case of the National Paediatric Hospital, Dublin, a project where costs have spiralled from the initial budget of €790m to a potential final account of €1.73bn as at April 2019 (PWC, 2019) and may well rise further as the project reaches completion. As the project stands, the cost increase is in the order of 220 per cent and, as such, would be seen as an indicator of major problems. Although this could be accounted for by major changes in project scope, such as those observed at the Scottish Parliament Building in Edinburgh, given the scale of the development and likely challenges such a development would pose. Equally, you could ask, was this down to estimator error? If it was, was this a deliberate misrepresentation as opined more generally in the work of Flyberg et al. (2004) to ease the political process and ensure the project initially proposed in 1993 actually became a reality? To answer this question, we have to look at the independent report produced by PricewaterhouseCoopers for the Irish Health Service Executive (PwC, 2019), which reveals major failings in project governance and financial management but also alludes to some area of possible unethical practice.

The PwC report reveals the timeline of the overspend, it is noted the overspend had not occurred all at once but was spread over various project milestones, as depicted in Figure 7.1.

From PwC's analysis of the initial overspend from the Developed Business Case until PwC's review in December 2018, additional expenditure of some £450m can be attributed to a series of events, including:

- *under-estimation* – there are some identified ethical concerns in the approach to project estimation and costing this accounts for €294m of overspend.
- *execution* – governance failings when developing the Guaranteed Maximum Price (GMP) accounted for €56m.
- *consequential costs* – costs attributable to other events accounted for €64m.
- *uncontrollable costs* – costs for which the client had no control accounted for €16m.

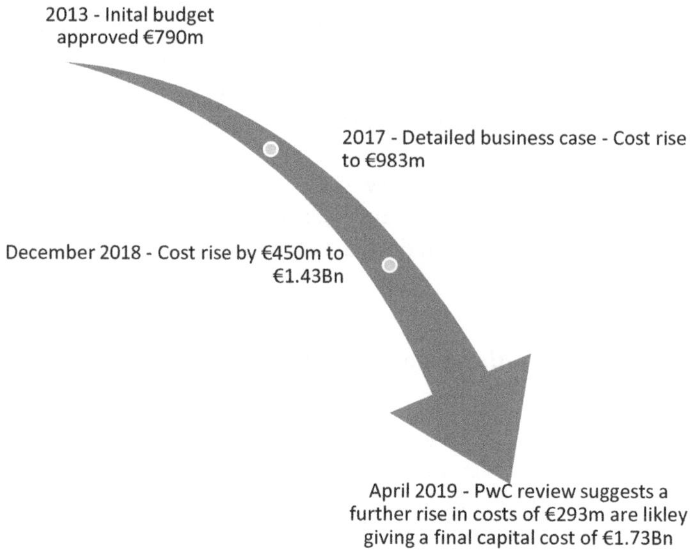

Figure 7.1 Timescales of the overspend
Source: Adapted from PwC (2019).

- *unclassified* – costs that cannot be directly proportioned to one specific category accounted for the final €20.4m.

While all cost centres are important, under-estimation is a major contributor to the full cost overrun with €294m or 65 per cent attributed to this cost centre and is also directly relevant to the ethical situation under review here. In the review of the project undertaken by PwC (2019, a number of issues emerged with the financial management of the project including:

1 Confusion about how the scope of works were quantified, with the consultant QS using a different approach to that adopted by the contractor.
2 Fragmented and inappropriate financial control systems for reporting financial updates, resulting in confusion.
3 Failure to follow the public sector spending codes when developing the business case.
4 Flawed budget development processes and poor risk management.
5 Significant under-estimation of budget at the point of agreeing the GMP.
6 Over-statement of cost certainty at design development stage and inaccurate reporting of design completion.

While some of these issues represent failures in project governance, there are also clear breaches of professional ethics in how quantity surveying processes have been instigated. The PwC report raises a number of concerns about the way in which the financial management process had been undertaken on the project that could raise ethical concerns about the practices implemented. The concerns raised centred around ineffective budget management, over-statement of design maturity and written assurances of inappropriate levels of cost certainty. Other issues raised in the report include a mismatch of measurement process, with the client team reliant on quantities developed from the BIM model whereas the contractors team developed quantities from traditional 2D paper drawings, leading to significant differences in quantities and thus pricing.

Discussion point 7.2

Having considered the ethics of project cost estimation using some hypothetical examples, how many of the issues identified by PwC (2019), relating to the National Paediatric Hospital, Dublin, do you feel constitute unethical practice and how many of them would you consider to be genuine errors due to failures in communication and project governance?

In this situation, there are in the main failings in communication and project governance, not significant breaches of ethical standards *per se*. The literature relating to project under-estimation (cost over-run) is well established, as summarised by Amadi and Higham (2019), with five principal explanations emerging:

1 The interplay of psychological factors on professional decision making.
2 Technical failings, errors or omissions by the professional team, such as lack of site investigations, challenges associated with project complexity and under-estimation due to various technical limitations.
3 Misestimation can also be due to the challenge of uncertainty and shifting project scope and economic actors.
4 The unethical practices, deception, delusion and the attitude of politicians and public officials.
5 Failings in project organisation and governance due to ineffective or inappropriate leadership.

It is acknowledged in the PwC report that a number of these factors have triggered the failings identified at the National Orthopaedic Hospital. Mapping the situation to the five issues identified from the literature, starting with psychological factors, the report clearly identifies issues with under-estimation due to technical failings, such as the adoption of two different ways of working when measuring the structure, the client team used the BIM model whereas the contractor adopted a more traditional measurement approach. Measurement error has been identified as a key contributor to cost over-run in the literature (Mackie and Preston, 1998) and contributes largely to the unintentional technical failings cluster of cost over-run. Second, the report raises the issue of optimism bias, this is a widely acknowledged phenomenon whereby the surveyor will be over-optimistic about the likely costs of various items of work. As such, the *HM Treasury Green Book* (HM Treasury, 2018) has identified a series of percentage increases for budgets to compensate for the effect of this. However, despite this guidance existing, the PwC report (PwC, 2019) suggests this percentage was not applied. This is a clear failing in the delivery of financial management on this project.

The report further raises the issues of a fragmented and inappropriate financial control systems for reporting financial updates, resulting in confusion and the general failure to follow the public sector spending codes when developing the business case. Both of these issues could be mapped to failings in project governance. However, they also raise ethical concerns around potential deception as more generally identified in the work of Flyberg et al. (2004). This, if taken with the observations of a deliberate over-statement of cost certainty at the design development stage and inaccurate reporting of design completion and the implementation of a flawed budget development process that sought to impede good governance, means it is clear this project has also suffered from issues that would raise ethical questions about the conduct of the team.

7.5 Theme 2 The financial management of projects at pre-design and design stages by the main contractor

7.5.1 Scenario 3: Not presenting the best solution to the client

In this scenario you are employed in the commercial team of a major contractor who has successfully bid for a D&B contract, but with a tight profit margin. The project is now well into the design phase and the design team have identified an issue with the proposed fire protection. The solution presented complies with all design standards and building regulations. Furthermore, the fire protection design achieves the minimum requirements of the performance specification. The issue is while the solution presents a good solution in terms of buildability, however, the fire engineers are concerned the solution proposed is not in the client's interests and could cause problems later on when the building is in use. While the solution enhances buildability, the solution has also allowed the contractor to implement value engineering on the project, thus improving their financial position.

? *Question 5*

The solution proposed is legally unproblematic with all minimum standards met and the contractual requirements of the performance specification met. As a result, the design will receive full statutory sign-off. However, the fire engineer feels the design could be enhanced to better achieve the client's requirements, but this would cause an increase in both the construction programme and project costing. How would you proceed?

Answer 5

This is a challenging decision for the team to make, on the one hand, there is no direct benefit to implementing the solution, contractually, this is not required as performance requirements have been attained. However, going beyond the minimum would enhance the value of the finished project from the perspective of the client. Ultimately this becomes a question of ethics and as a reader of this book, what do you feel the correct course of action to be?

Discussion point 7.3

Having considered the issues in the ethical problem, what do you think the correct course of action is? Do you adopt the most efficient process for the contracting organisation or do you act in the interest of your client?

Discussion point 7.3 is likely to have caused some debate, but this can be resolved with reference to the professional ethical frameworks of both the RICS and the CIOB. The RICS Global Professional and Ethical Standards outline five fundamental ethical obligations, several of which apply to this scenario.

First, 'act with integrity'. There are two key strands of this ethical standard that become relevant to this situation, first, the obligation to be open and transparent in the way you work and, second, to share appropriate and necessary information with your clients. So, one possible route out of this impasse is to propose the solution to the client as a potential design change (variation/compensation event) and to discuss openly the time and cost impact of enhancing the design. Alternatively, you can consider the second obligation of acting with integrity which is not to take advantage of your client as you owe them a duty of care. Does this duty of care extend to advising on potential better options? Yes, it does, especially as there is a perceived benefit to the client if this amendment were implemented.

The second ethical obligation relevant to this problem is the duty to 'always provide a high standard of service' within this ethical standard, the RICS standards espouse the importance of ensuring your client always receives 'the best possible advice, support or performance of the terms of engagement you have agreed'. This is also reflected in the third ethical standard 'Act in a way that promotes trust in the profession' (RICS, 2020). There is no expectation within the moral and ethical framework in which construction professionals operate for them to adopt a philanthropic approach to the management of their client's interests, however, there is the need to ensure the client receives the best possible service and alongside this, the client must be able to trust their project team, including the contractor. This would be especially important if this project had been let via an

NEC standard form of contract, where trust and co-operation and thus adoption of an ethical approach to business are a key obligation for both the employer (client) and contractor. So once again, despite the conditions of contract and performance requirements being achieved, referring this matter back to the client, highlighting the potential cost and time impacts of the proposal and seeking their consent via a variation would be the most appropriate way forward.

7.5.2 Scenario 4: Using the cheapest solutions to enhance profitability

Your firm has recently successfully secured a D&B contract, based on an NEC option A contract in a difficult market, which has forced down margins and tender prices, As a result, the procurement of the design team has been aggressively price-led. As a consequence, the design team are focusing on speed in an attempt to deliver the project within their professional fee tender figure. During the design process this focus on speed has caused a dispute between the structural engineer who has provided a solution for the frame of the building that fits within the cost envelope and the architect, who is complaining that the design puts severe constraints on the architect's room for manoeuvre and he has asked if some of the frame design could be amended. In attempting to resolve this dispute within the design team, it has become apparent that a better design solution exists, that has been used by the architect on a previous project, but due to the extra work and no perceived benefit to themselves, the structural engineer has refused to engage in any discussions with the architect. As the project manager for the contractor, how do you proceed?

Discussion point 7.4

Do you feel the structural engineer's approach is contractually appropriate? Furthermore, do you feel this is ethical? How would you then resolve this dispute?

In responding to this scenario, a number of questions emerge that the project manager would need to address. These will now form the basis for the following questions.

? *Question 6*

Is the structural engineer contractually correct?

Answer 6

Although this is not a legal textbook, this is an important starting point in developing a solution to this question. We do not have a copy of the professional services contract here, but generally there will be clauses within the contract outlining the obligations on the various parties. Given the proposed design meets the performance required and it is essentially designed with reasonable skill and care, in other words, the design complies with the usual practice and the professional standards applicable at the time of the design (Buckingham, n.d.), then the design would satisfy the contract.

? *Question 7*

Has the structural engineer behaved ethically?

Answer 7

There is quite some distance between morality and legal doctrine as explored earlier in this book. While the legal test has been satisfied, has the moral test? Looking to the ethics of this problem, the structural engineer has produced a design, that design has caused major complications for the design of the full scheme. The engineer is aware that a better design option exists, but due to the tight fee they are receiving, they are unwilling to amend the designs due to the extra work this creates and also the lack of direct benefit to themselves.

This situation does raise ethical questions, and, as a professional, this approach would not map well with the ethics of the professional bodies. Looking first to the CIOB Rules and Regulations of Professional Competence and Conduct, it is clear that in such situations, some of the ethical requirements are met, for instance, regulation 5.4 that requires the professional to ensure that any work they complete is 'executed in accordance with good practice and current standards and complies with all statutory and contractual requirements' (CIOB, 1993). This does not necessarily mean the approach adopted fully meets with all ethical requirements. Looking to the RICS, the first ethical standard is to 'act with integrity', has the structural engineer done that in this case? No, they have clearly acted in their own self-interest and not in the best interests of their client, who would ultimately receive an inferior building due to the severe constraints the current frame design has created. Furthermore, the engineer should be 'acting in a way that promotes trust in the profession', does this emerge in this example? Would refusing to amend the design to enhance the quality of the finished building or the prioritisation of the structural engineer's own interests achieve anything in terms of enhancing trust? The answer to both these question is no, so, once again, the actions are found to breach professional ethical standards. The final two ethical standards outlined in the RICS' Global Professional and Ethical Standards are: (1) to treat others with respect; and (2) to take responsibility. In terms of the former, it is clear from the dispute that the relationship with the architect is strained but also that the engineer is not treating their fellow professional with the level of respect the wider public would expect. Furthermore, in seeking a resolution, the engineer should take responsibility for the situation, look to rebuild trust with both the architect, the main contractor and ultimately the client and work with them to develop a workable solution. This may also have longer-term business benefits in terms of securing future work.

7.5.3 The case of the Grenfell Tower fire tragedy, London

The tragic events of 13 June 2017, which led to the loss of 72 lives in the Grenfell Tower fire, have been widely debated and with various public enquiries, criminal cases and other processes underway, we are yet to fully understand the events that led up to this tragic event. However, the ethics surrounding the events of that night have now become a key focus in the academic and wider literature. It therefore seems appropriate to include this event as a key ethical example in this chapter of the book. While the full scope of the events that led to this tragedy will not be explored, this case study will look at some of the issues associated with the design of the replacement cladding and the actions of stakeholders in the procurement of these works. A detailed account of the issues identified to date with the Grenfell disaster has been published by Dan Bulley and James Brassett, writing in the journal *Globalisations* (Bulley and Brassett, 2020). As such. this section of the book draws on this account, although with the latest updates from the inquiry as documented in *Building Magazine* at the time of writing in late 2020 to outline the facts of the case before exploring the ethical issues identified thus far.

7.5.3.1 Project background

While the fire at first appears to have started from the most mundane circumstances, a malfunctioning fridge freezer, much of the reason the disaster occurred has been linked to the 2015/2016 refurbishment of Grenfell Tower, and more specifically the specification and installation of an insulated cladding system to the building's façade and the concerns residents raised about the structure both before, during and after the completion of this refurbishment (Grenfell Action Group, 2016), with much of the criticism for the events at Grenfell focusing on the Royal Borough of Kensington and Chelsea Council (RBKCC) (Bulley and Brassett, 2020). However, the complex management structure in place at the time of the works has meant identifying the 'client' for this project has been difficult. While legally the building is owned by the Local Authority, they managed neither the building nor its renovation. Day-to-day responsibility for the asset rested primarily with a private not-for-profit management organisation, the Kensington and Chelsea Tenant Management Organisation (KCTMO). As a consequence of the divided responsibility between the management organisation and the building owner, it was determined that RBKCC would monitor the organisation's performance and approve all works over £400,000. Thus, governance and oversight for the Grenfell project included both organisations, but the day-to-day management of the project, including the competitive tendering of the refurbishment, rested with KCTMO.

7.5.3.2 The emergent ethical issues

While the fog that covers the Grenfell project continues to clear slowly piece by piece from the public enquiry, as of October 2020, a number of issues have been revealed that have raised legal and ethical concerns. This book is only interested in the latter, and so the issues around fire regulation and gross negligence manslaughter are not within the scope of this account. However, from the ethical perspective, a number of issues have emerged from both the initial report of the Grenfell inquiry and also from Dame Judith Hackitt's review.

In terms of the execution of project, it has been revealed that KCTMO took the lead, contracting the job out on a competitive basis, in line with the compulsory competitive tendering (CCT) policy aims which prized cost and efficiency. KCTMO ultimately appointed several firms to carry out the work (Moore-Bick, 2019, p. 3). This included the French-owned global project management firm, Artelia, with its professed specialism in cost management. Studio E was appointed as the architect, a London-based firm, boasting a portfolio of projects around the world. The lead contractor, however, was Rydon – selected because, unlike the initial contractors approached, it submitted 'the most economically advantageous tender', according to RBKC papers from 2014 (Bulley and Brassett, 2020). As is normal practice in construction, Rydon themselves subsequently subcontracted major aspects of the refurbishment project to a range of specialist contractors Cohen (2018), and identifying detectives are said to have been investigating as many of 500 companies by early 2018. However, the focus of the disaster has been the cladding, with the public inquiry's first phase report espousing this to be the 'principal reason' for the fire's rapid and deadly spread (Moore-Bick, 2019, p. 557). This case study will focus on the ethics surrounding the cladding of the building. The external cladding package, estimated to be worth circa £2.5m, had been let to Harley Facades Ltd. Although the actual process of subcontracting has not been disclosed, custom and practice within the construction market would suggest this was let in competition with a strong emphasis on price. The cladding used on the Grenfell Tower project consisted of aluminium composite material (ACM) panels manufactured by Ominis with a polyethylene insulation board substrate manufactured by Celotex, the RS5000 board (Bulley and Brassett, 2020). Both these cladding materials were then mechanically fixed to the existing concrete frame and brick infill cladding. The controversy in this specification includes: the fire testing process that deemed ACM panels safe, the lack of certification for the insulation substrate and finally the use of value engineering which changed the ACM cladding specification from Reynobond FR (Fire Resistant) to Reynobond PE (polyethylene) resulting in a saving of £2 per square metre (ibid.). Official documentation revealed that the original contract had included fire-retardant zinc panels, recommended by Studio E in 2012 and approved by residents. However, after KCTMO requested cost-saving measures from Artelia, these were replaced in 2014 by the cheaper ACM panels which saved nearly £300,000. This was undertaken alongside other value engineering interventions designed to reduce Rydon's initial winning bid for the project from £9.2 million, which importantly was already some £700,000 below the next lowest bidder and £1.2m below the highest tender, to £8.5 million which would allow the bid to meet the budget the management organisation had determined (Bulley and Brassett, 2020; Price, 2020; Dunton, 2020). As the Hackitt Review espoused, value engineering was used in this situation as a quick fix to resolve the budget deficit, value engineering in this case was about 'anything but value. It is cutting costs and quality' to make the project viable.

Why the management organisation became so strongly focused on delivering the project as cheaply as possible has been identified by Bulley and Brassett (2020) whose account of this budget-tightening process suggests:

> wider political decisions of imposed austerity associated with the global financial crisis in the UK had led to the coalition government inflicting a cap on local government borrowing in 2012. The low budget of nearly £10 million for the refurbishment was thus largely financed by the sale of RBKC-owned basements in Fulham which generated £8 million. This was partly why KCTMO turned to Rydon's cut-price bid.

In recent evidence to the public enquiry, it has now emerged that Rydon was approached about the need for value engineering even before the outcome of the tender exercise had been disclosed (Dunton, 2020).

While the public inquiry into this disaster continues and new and important revelations continue to emerge, it is felt it would be remiss of the authors to provide an opinion without knowing the full facts, but rather we wished to present this case study with a view of the future and to those readers of the textbook who may well wish to develop an opinion on the ethics of Grenfell we would like to end this case study with a discussion point. What are your thoughts on what you have read, both here and elsewhere on the ethics of Grenfell?

Discussion point 7.5

One of the emergent findings from the Hackitt Review, which focused on the fire regulation and safety of all stakeholders, was the need to 'put a focus on the way in which buildings are procured. If we have a process that makes people bid at cost they can't afford to delivery at, we set ourselves up to fail.'

Do you feel this is likely to change? Do you feel the legacy of Grenfell will be a root and branch reform of how buildings are procured? Do you feel clients will instead of making the project fit the budget at all costs start, in the words of Dame Judith Hackitt, to 'think about buildings not as jigsaw puzzles that magically come together' but start to treat them as complex system and look at the impact of all changes on the safety of not only the workforce but also users (be they residents or other stakeholders) and the wider public?

Finally, where do you feel the ethical and moral failings of the Grenfell Tower disaster rest?

7.6 Theme 3 Wider ethical issues from the perspective of both consultants and main contractors

So far in this chapter, key ethical challenges associated with the pre-design and design phase management of the project have been explored, often with a strong financial focus, given the role of the surveyor and project manager at this stage in the project lifecycle. Yet, it is also important to consider some of the wider ethical challenges these phases of the RIBA Plan of Work could reveal. As such, this section explores a wider range of real-life ethical problems.

7.6.1 Scenario 5: Tender document development

You have been tasked with developing the tender documents for a Design and Build project, the architect and MEP engineers have both encountered delay and as a result you have been left with one week in which to prepare the documents. Also, the tight fee award due to errors in the development of the fee bid for this project means your practice has only allocated 20 hours for this activity. Concerned both fee and deadline are unrealistic, you have challenged this, only to be told to 'do the best you can' and 'we make additional fee income from change as part of our contract with the client'.

? *Question 8*

How do you proceed? Would you book the hours to another project, try to complete the job in the timescale, knowing the documents are incomplete or request a more realistic time allocation and some help to achieve the tight turnaround?

Answer 8

Over the past 30 years, the pre-design and design phase of most projects has been accelerated, often associated with wider adoption of design and build and the removal of fee controls for consultants toward the increased adoption of competitive tendering for professional services. Those working in the sector have observed increasingly tightened delivery timescales and reductions in the time available to develop tender documentation. For some practices this has necessitated innovation, but for others this has led to some unethical working practices. In this scenario, the surveyor has effectively been asked to spend the allocated time and then issue the documents, but not to worry that they will be insufficiently developed as the practice will make more fee on the change events that will follow. This is obviously a completely inappropriate answer to the issue, regardless of the bid submitted, both the practice and individual surveyor have an ethical duty of care to their client. As part of this duty of care, the practice and thus the individual surveyor should both be acting with integrity and thus not use the situation to take advantage of the client to knowingly issue incomplete tender documents because they have bid in error for the work. Regardless of this, the practice and the surveyor should both ensure they 'always provide a high standard of work' (RICS, 2020). In the bid document the practice would have been clear about what service they intended to provide, and this must be honoured in full, regardless of the loss the practice would make. The RICS ethical codes mandate that the practice places 'the fair treatment of clients at the centre of its business culture' (ibid.) yet this is lacking here. Yet this is also an obligation on the individual to act ethically, whom the RICS codes (ibid.) stipulate 'must encourage their firm to do so', so the surveyor should raise this issue with their line manager. In this instance, the firm is putting their own commercial interests and greed ahead of the interests of the client by assuming any omissions in the documents would be helpful in attracting additional fee income. This is ethically unacceptable. The same argument would apply, and the same ethical framework would be relevant if the surveyor's line manager suggested booking the additional hours to a larger project within the practice. Although difficult to do, the surveyor should ask their line manager for a more realistic time allocation and some help to achieve the tight turnaround.

7.6.2 Scenario 6: Taking the client to the football

Your organisation is looking to tender for a design and build project with a new private client. During the initial invitation to tender stage, the new client let it be known that they really like to attend Premier League football. Sensing an opportunity, your commercial director has suggested a small team from the firm along with some of the client's staff attend the next home game of the local Premier League club and has already mentioned to the client the firm has a box at the ground.

? *Question 9*

If you worked for the client, would you accept this offer? Would your love of football influence your response?

Answer 9

Hopefully your love of football can be put to one side. Professionally the RICS advises that a construction professional should not 'accept directly or indirectly, anything that could constitute a bribe' (RICS, 2019a) with a bribe described in the same document as 'the offer, promise, giving, demanding or acceptance of an advantage as an inducement for an action that is illegal, unethical or a breach of trust'. In this instance, the invitation to the football clearly fulfils the definition above as you are in the process of tendering a major design and build project and the invitation is loaded with the hope of applying undue influence. As such, your acceptance of this would be a clear breach of ethical standards, as the CIOB's Rules and Regulations of Professional Competence and Conduct. make expressly clear 'at no time improperly offer or accept gifts or favours which would be interpreted by the Institute as exerting an influence to obtain preferential treatment' (CIOB, 1993), so this offer would be a clear breach of this ethical obligation.

? *Question 10*

Given the client has suggested this, does this mean your firm had to respond with the offer?

Answer 10

This is more difficult; the client has insinuated they are aware that your organisation has a box at the football, and they have very discreetly suggested they would like an invitation. They have not promised this would influence the outcome of the tender process, but the insinuation is clearly there. The benefits of making the offer are very clear. However, the legal and ethical implications of this are equally clear. First, as discussed extensively in Chapter 4 of this book, this process would fall under the legislative envelope of the Bribery Act (2010), both in terms of Section 1, as the contractor in this situation is willing to extend an invitation of the company box in full knowledge this is likely to at least partially influence the outcome of the tender process. Furthermore, the client would be equally impacted under Section 2 as they have very clearly sought and are equally willing to thus accept an invitation to the football, which, on the balance of probabilities, is likely to be positively received and thus could be construed as influencing the outcome of the tender process. As well as the legal implications outlined, there are also clear breaches of ethics, the RICS Rules of Conduct for Firms (RICS, 2020) make it clear that 'A Firm shall at all times act with integrity and avoid conflicts of interest and avoid any actions or situations that are inconsistent with its professional obligations.' While the CIOB's Code of Professional Conduct and Rules similarly asserts that firms should manage 'their affairs so that all its actions are conducted in accordance with good business practice' (CIOB, 2015). So for both organisations, this situation presents a clear breach of legal-ethical standards.

7.6.3 Scenario 7: Managing client money: payment of fees upfront

One of the commonest ethical dilemmas encountered in practice is the management of client money, this is an issue often explored during the RICS APC interview, and which seems to cause some confusion among those answering it. So, the scenario below seeks to address this point head on.

? *Question 11*

You have been reappointed by a long-term, overseas client to act as employer's agent and contract administrator for a series of new developments in the

United Kingdom. Given the challenges they have encountered in the past with international fund transfers, they have asked if they could place the working capital for the initial stage of the development in your account, allowing you to draw down the funds as needed for the interim payments. Do you feel this is acceptable?

Answer 11

This situation opens up several legal and ethical concerns that must be addressed. Starting with the legal concerns, the key issue here is the legitimacy of the funds in question. As the work is based in the UK, the company would first need to ensure the Money Laundering and Terrorist Financing (Amendments) Regulations 2019 have been fully complied with, specifically, as the transaction would include the movement of funds over €10,000 so the organisation would need to undertake customer due diligence (Pinsent Masons, 2020). The compliance requirements the business would need to engage with under customer due diligence (CDD) include the examination of the background of the business and purpose of the business relationships and transactions with customers in high risk countries (ibid.). Additionally, advice from the Financial Conduct Authority (2020) identifies that the firm would (under Regulation 28) need to understand the ownership and control structures within the corporate customer's organisation, check any such information against anything held at Companies House if the corporate client has any UK businesses in their direct control.

Once the legal governance obligations have been satisfied and the money can legally be accepted, the next issue to consider is the ethics of this. From the ethical perspective, there are two approaches the firm could explore. The first option is to look at project bank accounts (PBA). Project bank accounts are ringfenced bank accounts set up with the sole purpose of channelling payments within a construction project to the main contractor and subcontractors. The employer (the practice's client in this instance) is able to maintain funds in the account several months in advance, while also ensuring fair payment practices are maintained for project stakeholders (Fenwick Elliot, 2012). This approach is considered ethically to be at the forefront of best practice, aligning well with the UK government's commitments outlined in the Government Construction Strategy. The project bank account can easily be incorporated into the construction contract for the project, for example, if this was let using the Joint Contracts Tribunal (JCT) 2016 suite, the PBA supplement could be adopted. The supplement includes various enabling provisions to be incorporated into the JCT contract and a form of PBA (Dentons, 2019) although the client would incur additional costs associated with establishing and administrating the account.

The alternative is to deposit the funds with the practice. For this to happen, a series of legal and professional bodies' rules of conduct safeguards must be met. Before these safeguards are explored, the RICS Rules of Conduct express that a clear governance framework is required. Rule 8 of the RICS Rules of Conduct for Firms states 'A firm shall preserve the security of clients' money entrusted to its care in the course of its practice or business' (RICS, 2020). From the perspective of the RICS, assuming the firm is RICS regulated (which it is likely to be), the firm is required to first register with the 'RICS Client Money Protection Scheme for Surveying Services', thus ensuring compliance with the Client Money Protection Schemes for Property Agents (Approval and Designation of Schemes) Regulations 2018, before then implementing a robust governance, system in place for the handling and use of the client's funds. The details of this are outlined in the RICS guidance note 'Client Money Handling' (RICS, 2019b),, which advises that a clear and auditable process of fund management must be adopted, demonstrating receipt, storage and use of the funds to the client at all times.

7.6.4 Scenario 8: Conflict of interest: working as a freelancer

After several years working as a quantity surveyor, you decided to work freelance and set up your own micro business offering quantity surveying services to both clients and main contractors, both directly and through employment agencies. During the course of these activities you have developed working relationships with several clients and contractors, often managing the pre-contract processes, usually based on contract drafting. You have recently drafted the prelims and supporting contract provisions for a small project for a new client which has now been sent out to tender. This morning a regular client has telephoned asking your advice about a contract they are currently tendering for. As the conversation developed, you realised this is the project you drafted the contract for, you are now worried that refusing to offer the advice would lose you future work with this client.

? *Question 12*

How do you proceed in this situation?

Answer 12

This is a difficult situation to have found yourself in but a very common one when working as a freelance construction professional. In theory, you are entitled to work for whoever you choose and accept offers of employment when they become available. It is also not uncommon for larger practices to work for multiple parties, which is the situation outlined in the scenario here. However, there is also a very clear conflict of interest created. The RICS describe conflicts of interest as 'a situation in which the duty of an RICS member … to act in the interests of client … conflicts with a duty owed to another client' (RICS, 2017). In the above scenario, if you proceeded to accept the work from your regular client, you would become employed by both the regular client and the client tendering the project. Based on the contractual advice you gave, this clearly means you have been and remain privy to confidential information and have an awareness of how a commercial advantage could be gained by those companies wishing to submit a tender. If you then use this information for the benefit of the main contractor, this will be operating with knowledge of a conflict of interest and proceeding ahead regardless of your own financial benefit.

While it may appear unfortunate and against the business 'idea', as a freelance consultant, especially as the offer of this work has come from a regular client, your relationship with this client could now be jeopardised. However, ethically it is also clear that you have an ethical duty to your client to act in their best interests. So clearly advising both parties would construct a breach of this ethical duty. Although this would not be illegal and so you could proceed, as discussed throughout this book, ethics go above and beyond what is required by law. The RICS guidance does go further and advises that if you are considering proceeding with an offer that may cause a conflict of interest, then discussions should be had with all relevant parties so they are aware of the potential situation and can provide feedback (ibid.). Rule 5 of the CIOB Rules and Regulations of Professional Competence and Conduct, also state that 'members shall discharge their duties with complete fidelity and probity'. In particular, the following rules apply:

- Rule 5.1: Not divulge to any person, firm or company any information of a confidential nature relating to the business activities or processes of their employer or client acquired during the course of their work.
- Rule 5.2: Not, without the permission of their employer or client, render any service, with or without remuneration, which conflicts with the interests of their employer or client.
- Rule 5.3: Ensure, when providing an advisory service, that the advice given is fair and unbiased.

Therefore it is clear, that if the work was accepted, clear parameters would have to be drawn up between yourself and the main contractor about what exactly you are expected to consult on, and on which areas you are unable to offer advice, in order to avoid and minimise any conflict of interest where possible. Both the contractor and client would also need to agree to these actions, which in this situation would be unlikely.

As a more realistic approach, as a micro business, you would be advised to turn down the work but outline the ethical reasons for doing so, this your client will surely respect and thus it will not sour the relationship. The alternative would be to proceed with this work, but to employ a second freelance professional and establish a 'Chinese wall' between you and them to maintain client confidentiality. The concept of a Chinese wall has been defined by McKenna (1998) as:

> A Chinese wall is an internal measure adopted by a firm to ensure that information gained while acting for one client does not leak to people in

another part of the same firm who are acting for another client to whom that information may be highly relevant.

Ultimately, it is important to ensure no conflict of interest exists, it is only by not having a conflict of interest in place, either by not accepting a position that comes with a conflict of interest, or by eliminating one that exists. This is how a professional can meet the RICS ethical standards of 'acting with integrity', 'Always providing a high standard of service' and 'Act in a way that promotes trust in the profession' (RICS, 2020).

7.7 Summary

This chapter has explored how ethical frameworks can be applied to real-world construction problems that could occur in the early stages of a project, while also helping you and your fellow construction professionals make better decisions in the face of difficult situations. The chapter has also looked at two real-life examples to demonstrate how these ethical challenges can play out in the realities of construction sector. There is without a doubt a difference between the presentation of real ethical situations when are they responded to in a theoretical discussion, with the benefits of hindsight, time and research to reach correct responses. The realities of these in professional life, with the pressure of time applied to the decision, will often be different, actually in the comprehensiveness of the response but not in the key ethical positions adopted. It is hoped this review of some of the myriad of ethical issues that arise in the life cycle of a construction project will at least assist you, the construction professional, and aid understanding of how these issues can be addressed competently and ethically. It is hoped these examples will also help those seeking chartership who, when in a high-pressure situation, are presented with questions of ethical judgement. The key take-away message from this chapter is ethical dilemmas are challenging, complex and often have no single answer. It is our duty as professionals to have a thorough understanding of professional ethics and to be able to apply these to the situation we face. Ultimately, as we do in our wider life, you will develop an intuition for the correct ethical decision and will implement this often without giving it a second thought. Regrettably, however, not all professionals adhere to these ethical norms of behaviour and the ultimate test becomes one of legal responsibility under the tort of negligence or the view of the panel in an employment or professional body hearing.

References

Amadi, A.I. and Higham, A.P. (2019) .Putting context to numbers: A geotechnical risk trajectory to cost overrun extremism. *Construction Management and Economics*, 37(4): 217–237.

Buckingham, S. (n.d.) Understanding your design duty – "Reasonable skill and care" vs. "fitness for purpose" – mutually incompatible or comfortably coexistent. Available at:

www.fenwickelliott.com/research-insight/annual-review/2014/understanding-design-duty

Bulley, D. and Brassett, J. (2020). Everyday ethics of the global event: Grenfell Tower and the politics of responsibility. *Globalizations*. Available at: www.tandfonline.com/doi/full/10.1080/14747731.2020.1798109

Cantarelli, C.C., Flyberg, B. and Molin, E.J.E. (2010). Cost overruns in large-scale transportation infrastructure projects: Explanations and their theoretical embeddedness. *European Journal of Transport and Infrastructure Research*, 10(1): 5–18.

CIB (Chartered Institute of Building) (1993). Rules and Regulations of Professional Competence and Conduct. Available at: www.ciob.org/sites/default/files/Rules%20&%20Regulations_0.pdf

CIB (Chartered Institute of Building) (2015). Chartered Building Consultancy: Professionalism and integrity in construction. Available at: www.ciob.org/sites/default/files/Code%20of%20Conduct%20for%20Chartered%20Building%20Consultancies_0.pdf

Cohen, D. (2018). Leaked Grenfell dossier reveals how disastrous refurbishment turned tower into a 'tinderbox'. *Evening Standard*. Available at: www.standard.co.uk/news/london/shock-grenfell-dossier-revealsdisastrous-refurbishment-turned-tower-into-a-tinderbox-a3814866.html

Dentons (2019). Project Bank Accounts: Making payment fair. Available at: www.dentons.com/en/insights/articles/2019/february/27/project-bank-accounts-making-payment-fair

Dunton, J. (2020). Grenfell Inquiry: KCTMO's second week in the spotlight. Available at: www.building.co.uk/news/grenfell-inquiry-kctmos-second-week-in-the-spotlight/5108637.article

Fenwick Elliott (2012). Project Bank accounts: The way forward? Available at: www.fenwickelliott.com/sites/default/files/insight_issue_13.pdf

Financial Conduct Authority (2020). Money Laundering Regulations. Available at: www.fca.org.uk/firms/financial-crime/money-laundering-regulations.

Flyberg, B., Skamris Holm, M.K. and Buhl, S.L. (2004). What causes cost overrun in transport infrastructure projects? *Transport Reviews*, 24(1): 3–18.

Grenfell Action Group (2016). KCTMO: Playing with fire! Available at: https://grenfellactiongroup.wordpress.com/2016/11/20/kctmo-playing-with-fire/

Hackitt, J. (2018) Building a safer future. Independent review of building regulations and fire safety. Available at: https://assets.publishing.service.gov.uk/government/uploads/system/uploads/attachment_data/file/707785/Building_a_Safer_Future_-_web.pdf

HM Treasury (2018). *The Green Book: Central Government Guidance on Appraisal and Evaluation*. London: HMSO.

Mackie, P. and Preston, J. (1998). Twenty-one sources of error and bias in transport project appraisal. *Transport Policy*, 5(1), 1–7.

McKenna, C. (1998). Chinese walls: Maintaining client confidentiality. Available at: https://uk.practicallaw.thomsonreuters.com/3-100-8763?__lrTS=20171228112637910&transitionType=Default&contextData=(sc.Default)&firstPage=true

Ministry of Housing, Communities and Local Government (2018). *Building a Safer Future. Independent Review of Building Regulations and Fire Safety: Final Report*. London: HMSO.

Moore-Bick, M. (2019). *Grenfell Tower Inquiry: Phase 1 Report*. London: HMSO.

NBS (2018). *National Construction Contracts and Law Report 2018*. Newcastle upon Tyne: RIBA Enterprises Ltd.

Pinsent Masons (2020). UK updates Anti-money Laundering Rules. Available at: www.pinsentmasons.com/out-law/news/uk-updates-anti-money-laundering-rules

Price, D. (2020). Rydon rivals' Grenfell bids revealed at inquiry. Available at: www.constructionnews.co.uk/news/rydon-rivals-grenfell-bids-revealed-at-inquiry-29-01-2020/

PwC (2019) .New Children's Hospital: Independent review of escalation in costs. Available at: https://merrionstreet.ie/MerrionStreet/en/News-Room/Releases/20190409_NCH_Report.pdf

RIBA (Royal Institute of British Architects) (2020). RIBA Plan of Work 2020. Available at: www.architecture.com/-/media/GatherContent/Test-resources-page/Additional-Documents/2020RIBAPlanofWorktemplatepdf.pdf

RICS (Royal Institution of Chartered Surveyors) (2017). *Conflicts of Interest.* Coventry: RICS.

RICS (Royal Institution of Chartered Surveyors) (2019a). *Countering Bribery and Corruption: Money Laundering and Terrorist Financing.* Coventry: RICS.

RICS (Royal Institution of Chartered Surveyors) (2019b). *RICS Professional Standards and Guidance: UK Client Money Handling.* Coventry: RICS.

RICS (Royal Institution of Chartered Surveyors) (2020). The Global Professional and Ethical Standards. Available at: www.rics.org/globalassets/rics-website/media/upholding-professional-standards/standards-of-conduct/the-global-professional-and-ethical-standards.pdf

8 Construction phase ethical dilemmas

8.1 Introduction

This chapter is concerned with the application of ethical frameworks in practical construction industry-based scenarios. There are five main sections: working for a main contractor, working for a subcontractor, being self-employed, working for a consultant or working for a client. While the construction industry is diverse and many professionals may consider themselves part of organisations that do not specifically meet these category headings, the headings are broad enough to cover the majority of professionals currently operating. Several ethical scenarios are provided, each with accompanying questions and answers about the actions required to be undertaken. The situations provided are themselves quite broad and can easily be experienced in any constriction industry role, not necessarily simply only by those professionals who are operating in that particular category. Answers are provided as to how such scenarios should be dealt with and how such questions could be answered, in the authors' opinions, with reference to standards and elements from different ethical frameworks. Sometimes an ethical framework can appear to be created in isolation from a real-world setting, and its application is difficult to imagine. This chapter attempts to address this issue and help foster an understanding of how ethical standards can be applied to many different situations, and their application clearly discussed to assist in evidencing a construction professional's knowledge of ethics.

8.2 RIBA Stage 5

The Royal Institute of British Architects (RIBA) is the leading professional body for architects, established in 1837. The aim of RIBA is to drive 'excellence in architecture'. In 2020, the latest RIBA Plan of Work was launched, comprising eight work stages. Stage 5 is 'Manufacturing and Construction' and it is during this stage the building is constructed on site, and any major construction element fabricated off site.

However, it is important to remember the RIBA Plan of Work is simply a framework through which RIBA envisages and recommends construction decisions are taken. Nevertheless, RIBA itself is a respected and industry leading body and the Plan of Work is an effective framework for discussing and focusing upon the key stages in a construction project. What most people see as 'construction work' occurs under this single stage. This chapter therefore focuses on the personal, organisational and professional ethics that could potentially arise during the construction phase of a typical project.

8.3 Professional ethics

The following are a series of real-life examples which pose ethical dilemmas that you may face during your professional career. You may consider some to be small ethical dilemmas where the course of action you should take is clear, others are more substantial and complex. It most cases, all possible courses of action have far-reaching consequences, both indirect and direct, immediate and in the future, that are not ideal, regardless of the choices you make. While so far this textbook has explored all manner of ethical frameworks from a somewhat theoretical perspective, this chapter takes these frameworks and applies them to a set of real-world scenarios to help you navigate the ethical dilemmas you may face or be asked to advise upon throughout your professional career.

Predominantly this chapter will focus on the ethical issues encountered by construction industry professionals during the execution of their daily responsibilities. For each example presented below ethical standards and frameworks will be applied as worked examples that will guide best practice behaviour in a variety of ethical scenarios.

8.4 You work for a main contractor

8.4.1 Scenario 1: Final account

You have responsibility for agreeing the final accounts with a series of subcontractors. The groundworks subcontractor had an original order value of £850,000. With agreed extras this has risen to £1.1 million. The groundworks subcontractor has also been applying in their application for payment for additional works of £200,000, taking the potential full final account value to £1.3 million. The £200,000 of extras do not have agreed variation instructions in place but you are aware that a large part, if not all parts, have been completed on site. You meet the groundworker's quantity surveyor site and walk around the project discussing the extra works.

? *Question 1*

You can see the additional £200,000 of extra works the groundworker is applying for has been completed, and all to a high quality. However, it is apparent that the groundworks subcontractor, although hard-working and proactive in accommodating on site requests and changes, is not competent in contract administration and management, and so has left themselves exposed commercially. They have completed the extras taking the value of their works carried out to £1.3 million but can only contractually claim for the £1.1 million they have already been paid. As the groundworks subcontractor cannot contractually claim for the additional works, their QS has asked you to pay them anyway as the work has been done. What are your first thoughts on this situation? Would you pay the £200,000?

Answer 1

You may immediately think of paying the additional works as you can see it has been completed, or you may immediately think of not paying the works as the subcontractor has not followed the procedures as set out in the contract. Both immediate thoughts would have correct elements. After all, there is a contract in place to avoid situations like this from occurring, and if one party has not followed the contract correctly, then they are at fault and so you may subscribe to the opinion that as the groundworker has not followed the contract in regards to the extras they are applying for, then they should not be entitled to any of the sums claimed.

However, from an ethical perspective, is this correct? If we consider the first of the five RICS Global Professional and Ethical Standards (RICS, 2020a) 'Act with integrity', can we describe a construction industry professional who purposefully does not pay another company for work that has been carried out as acting with integrity? There is often a misunderstanding when it comes to the administration of contracts. While it is obviously in everyone's interests to follow the agreements as outlined in the signed contracts, a recent construction industry report revealed that 33 per cent of construction contracts are either signed after work has already begun or never signed at all (NBS, 2015). Therefore, if this is an instance where a contract has not yet been signed, this may be because it has not yet been reviewed, or has issues outstanding but the subcontractor has started work in good faith before the signing of the contract at the request of the main contractor (a common occurrence in the industry). Therefore, the contractor may not have been aware of the procedure to follow in all instances.

Also, it is important to discuss the spirit of a contract versus the wording. The spirit of every construction contract is to agree that works will be carried out by one party to a high standard, and in exchange the other party will pay any agreed sum. The wording of the contract is the manner in which all envisaged eventualities are captured and agreed. Therefore, if a subcontractor has carried out all the work to a high standard (this will of course need confirming with the site team), as requested and required, they should be paid fairly for that work in accordance with the spirit of the contract, even if the wording of the contract allows 'wiggle room' for the main contractor to argue otherwise. Maximising this 'wiggle room' at the expense of the subcontractor is not acting with integrity.

ⓘ *Question 2*

After investigating this further with the construction manager, it appears the work was verbally instructed on site. The subcontractors proceeded with the works as they were told as it was a critical element in the project's completion. The subcontractor has carried out the additional aspects of work in good faith with the site team, who may not have realised payment would be an issue and presumed all works would be paid for. How would you proceed?

Answer 2

The RICS Global Professional and Ethical Standard of 'Act in a way that promotes trust in the profession' (RICS, 2020a) could be understood in this context. While the contract procedures have still not been followed correctly, this is the fault of both the main contractor's site team and the subcontractor and, therefore, it would be incorrect to penalise one side financially when both parties have not followed the agreed contract procedures.

Trust extends to all stakeholders, including the party who is contracted to fulfil the works. They should be able to trust you, as the main contractor's employee, to treat them fairly and abide by the highest standards. As a professional you should also be aware of the findings of a recent industry report which stated that 48 per cent of contractors reported finding contracts difficult to understand (Bibby Financial Services, 2019). Even if a contract has been reviewed and signed prior to work commencing, the subcontractor may not be fully aware of all their requirements. While you can argue that this exposes the subcontractor to elements of risk and it is their responsibility to ensure they fully understand all contract obligations, you have to take into consideration the subcontractor's contract knowledge and ability may be different from their trade knowledge and ability.

ⓘ *Question 3*

After the discussion on site, the QS from the groundworks subcontractor states that the company really needs the £200,000, and although they believe they are entitled to it, they ask if you will certify the funds as 'it's not your money' and they will give you £5,000 cash. How would you react in this situation?

Answer 3

While the contractual and ethical obligations remain over the agreement and payment of any additional works as discussed in the answers to questions 1 and 2, the situation has now changed, and your reaction will need to adjust accordingly. It is not wise to ignore such comments, even if they are made in an informal and joking manner. Depending upon the context and your own experience and relationship with the subcontractor, you may wish to deal with this issue by politely refusing, and then stating your reasons as to why. It is then advisable to inform your line manager of the occurrence and explain how you have dealt with the issue. If no further discussion regarding a potential bribe is raised, then you should continue the final account valuation as normal. You will need to be conscious that despite such 'offers' from the subcontractor, you should still act with integrity and value their final account fairly and accurately in conjunction with the items discussions in Answers 1 and 2.

If the 'offer' from the subcontractor is a more serious one, in addition to the above suggestions, you should also inform the individual they would be guilty under Section 1 of the Bribery Act (2010) in bribing another person, and you would be guilty under Section 2 of accepting a bribe, and also their company could be found guilty under Section 7 of the Act of failing to prevent a bribe from being paid on their behalf. In such instances, and upon discussion with your line manager, a more formal response may be required from your organisation to the subcontractor, setting out what has happened on the project and clearly stating your position on a matter you consider to be attempted bribery. All future working relationships should also be considered in light of their response to your formal letter.

8.4.2 Scenario 2: Cladding procurement

You have been tasked with procuring several work packages and the task is taking a lot longer than you had anticipated. You are under pressure from your boss and generally not very happy at work at the moment. During the procurement of the cladding package you have three returned tenders, two from companies you have not used before and one from a company you have used previously. The cladding subcontractor you have used previously has returned the lowest cost, and so you decide to award the package to them.

⑦ *Question 4*

Is the decision you made the correct one? Are you correct to award simply on lowest cost and previous working relationships?

Answer 4

This depends on the procurement criteria set during the compilation of procurement documents and communicated out to all interested tenderers. Clear procurement criteria along with accurately described weightings for each element and robust works information will allow all parties to price the works fairly and accurately. All information communicated should be consistent, and the manner in which queries were dealt with should be transparent. Prior working relationships should not form a part of any award decision. However, experience of any potential supply chain partner can (and should) form part of the awarding criteria, and inevitably those you have worked with previously may score higher as you have first-hand experience of their competence.

Nevertheless, this should not be used as a barrier to prevent new companies being successful in their procurement attempts with your company.

To ensure fair and rigorous procurement is conducted, that can withstand any scrutiny over the final decision made, a well-known procurement framework should be adopted. This could, for example, be based on the 'five rights of purchasing' as proposed by CIPS (2015). These are the right price, quality, time, quantity and place. Such as framework is basic but covers all the key aspects that should be considered in procurement and will help a balanced and fair decision to be reached. Simply awarding on the lowest immediate cost may not be the most sensible decision when considering the wider tender against the requirements of the project in the long term.

Ⓘ Question 5

Before you make the final formal contract award, one of the cladding subcontractors who did not return the lowest cost contacts you privately and offers a final discount that takes their price 5 per cent lower than the previous lowest tender. Is it correct to now award the new subcontractor the works package?

Answer 5

According to RICS guidance, tender extensions are a normal procedure but if offered, they should be offered to all subcontractors submitting tenders to avoid any unfair advantages. However, late tenders should not be considered for public project projects, and private clients should be informed of any late tenders and their agreement and instruction should be sought prior to including such tenders in the final consideration (RICS, 2014). While this guidance is more aimed at consultants, it should still be followed by main contractors. If they are operating on a design and build procurement route, there may be less day-to-day scrutiny by the client of the procurement and tendering practices that are occurring. Nevertheless, clear tender rules need to be in place and communicated to all those subcontractors who express an interest in submitting a tender for any work packages during the issuing of project information. It is ethical in procurement to have clear rules for all parties, and to strictly adhere to these rules. That is not to say flexibility should be discouraged, but that any flexibility afforded to one party should be afforded to all parties equally.

In this instance, the offer should be rejected. While it may seem appealing to accept a lower price from a competitor, or it may seem equal and fair to go back to all parties and ask them for a Best And Final Offer (BAFO), such practices should be avoided. First, if you did decide to accept the revised offer, this would be unfairly penalising the other contractors who submitted tenders for playing by the rules you set. Second, any sort of request for a discount or submitted cost reduction is simply bad practice and is focused on the 'squeezing' of the supply chain at the expense of any value they may have in the project. If you have issued robust procurement documentation to a minimum of three contractors, and they have all had a suitable amount of time to price and return their tenders, then a contract award should be made based on the cost and tender details you have received. If the costs are not suitable, it is the amount and type of details issued as part of the procurement, the project requirements, or an unrealistic budget that are the issues to be addressed.

8.5 You work for a subcontractor

8.5.1 Scenario 3: Offer of a Christmas meal from a supplier

You are a site-based project manager working for a medium-sized mechanical and electrical subcontractor on a housing development project. You regularly sub-let elements of works and you have done this with the procurement of the CCTV work several months earlier. Works are progressing as expected on the project and as Christmas is approaching, the CCTV subcontractor has offered to take you and several members of the management team out for a meal.

ⓘ *Question 6*

As the project PM, do you accept the offer or decline?

Answer 6

According to the RICS, a professional must not 'accept directly or indirectly, anything that could constitute a bribe' (RICS, 2019) with a bribe described by the same document as 'the offer, promise, giving, demanding or acceptance of an advantage as an inducement for an action that is illegal, unethical or a breach of trust'. In this instance, the invitation to a meal does not initially fulfil the definition above.

Before you proceed, however, questions must be asked of the timing and cost of any offer such as this. If it is during or before a procurement decision is to be made, then it should be declined, and if the cost of the meal is deemed to be too high, then it should also be declined. However, in circumstances such as this, it is organisational policy that should be referred to, and any decision to accept should only be made after consultation of the applicable organisational policies and discussion with the organisational management. Such policy should include the need for all gifts and hospitality to be transparently recorded and communicated to all staff, so they are aware how to behave in similar circumstances. For instances such as this, all guidance is somewhat purposefully ambiguous as there are so many different factors that need to be considered. Ultimately it comes down to the organisational policy and procedures that are in place.

8.5.2 Scenario 4: Bid fixing

As an Estimating Director for a specialist subcontractor, you are responsible for securing new work. The subcontractor is so specialist that there are only two other companies who offer precisely what your company does, in the locations your company operates. You are approached by a Director of one of the other specialist subcontractors.

⑦ *Question 7*

The Director of your competitor raises the idea that as every main contractor who wants what your company offers usually goes to all three companies for a comparable price, maybe you should all discuss the prices you intend to submit prior to submission. This will allow two of you to artificially inflate your prices by 50 per cent and the third by 25 per cent, rotating who will submit the 25 per cent higher bid each time. Their thoughts are that this will help the profit margins of all three companies rather than constantly competing against one another. What do you do in this instance?

Answer 7

While this may at first appear tempting as the main contractor customers would have nowhere else to go for the specialist services you offer, and it would increase your profitability, such an agreement is illegal and ethically wrong. Both the Enterprise Act (2002) and the Competition Act (1998) prevent UK organisations abusing a dominant market position and entering into anti-competitive agreements with other organisations. If you entered into the agreement as suggested by the Director of your competitor, then you would fall foul of UK legislation and potentially be criminally liable, plus be open to a fine of up to 10 per cent of your global turnover and potentially liable to claims for damages from both customers and other competitors.

There are many real examples of such behaviour in industry, such as an investigation by the Competition and Markets Authority which revealed illegal cartel-like behaviour between three concrete drainage product suppliers, who, between them, were found guilty and faced fines of over £36m (Prior, 2019). Such behaviour would also be against international ethical principles, such as those proposed by the International Ethics Standards Coalition. For example, your financial dealings would not be truthful and trustworthy (Financial Responsibility Standard), would not be observing the legal requirements applicable (lawfulness) and would not be promoting the reputation of the industry (trust) (IES-C, 2016).

8.6 You work for a consultant

8.6.1 Scenario 5: Pressure to circumvent standards

You are a new employee to a consultancy that specialises in providing project management and construction services. You are tasked with assisting an experienced consultant with preparing a cost plan, including all taking off responsibilities. You both anticipate it will take you 20 workdays, but your line manager has said they want the finished cost plan in ten workdays.

? *Question 8*

Your experienced colleague has said this situation occurs often and the only way to successfully deliver the cost plan in time is by not following the standards set out in the RICS NRM, and your colleague says, 'It is okay as it is only advice the client needs at this stage and they will get an accurate price when it goes out to tender.' How do you proceed?

Answer 8

If you were to proceed with the above task, it is probable that failure will occur at some point. Either the project will be delivered on time but not to the quality expected, or that the required completion date will not be met. In either instance, you will not be providing a high standard of service which is the second RICS Global Professional and Ethical Standard (RICS, 2020a). The third standard of 'act with integrity' would also not be achieved if you were to proceed as you intended to circumvent professional standards, standards that the client expects you to adhere to. In this instance, you will need to have an open discussion with your colleague and management regarding the timeframe of the project. This could include a revised plan to allow each section of the cost plan to be released once it is ready, and all professional standards have been adhered to. Alternatively, a revised completion date for the entire project may be agreed. It is not advisable to proceed with the intention to circumvent standards as this could lead to incorrect advice being provided to the client and ultimately could lead to the withdrawal of an individual's professional membership.

? *Question 9*

After you have looked at the details of the brief, you realise this is not something you have any experience of and are not sure you can do what is required of you. How do you proceed?

Answer 9

The third RICS Global Professional and Ethical Standard is 'Always provide a high standard of service' and this standard includes the expected behaviour that all professionals should act within their competence (RICS, 2020a). Therefore, before you proceed with the brief, you need to ensure you fully understand the requirements and expectations of your line manager and colleague. As you are a new employee to the consultancy, the expectation may be that you observe and learn what to do for the future, and as long as all parties are clear on those expectations, it will be okay to proceed.

However, if the expectation is that you are to be a fully involved member of the team and contribute to the work, then you would not be able to do so in the manner expected. Therefore, you should not proceed, as any work you do contribute may ultimately slow the project down and/or lower the standard of service delivered to the client. In this instance, you should discuss the clear parameters of your role and expected contributions with all staff involved at the earliest stage possible.

8.6.2 Scenario 6: Suspicion of money laundering

As an experienced consultant, you have recently moved to a new company that is young but growing quickly. You have a middle management position and currently have responsibility for advising a new international client who is looking to invest in the UK property market. This is not your first time advising international clients, or those new to the construction and property industry, but you do not have vast experience in this area and have not worked previously with clients from this part of the world.

ⓘ *Question 10*

Your new client has lots of money to spend, you are not sure exactly how much but it is in the range of several millions, and they are willing to transfer large amounts to your company to cover all professional fees, to be held in a client account from which you can purchase property and land and hire further consultants and contractors on their behalf for any construction works required. However, they are not forthcoming on the source of their funds and have ignored your two requests on this matter to date. They want to transfer the funds immediately and start buying property and land. Do you proceed?

Answer 10

RICS guidance is clear that regulated firms should not 'facilitate or be complicit in money laundering' (RICS, 2019). In the first instance, such refusals to answer your queries on the source of any funds should act as a red flag: 'common characteristics that either individually or in combination might indicate potential misuse of the real estate sector for money laundering or terrorist financing purposes' (ibid.).

Even though you are new to the company and may not be familiar with all the processes in place, any suspicions should be reported to a superior. You will need to check what procedures the organisation has in place and ensure these are followed. All organisations that deal with client funds should have procedures and safeguards in place for such occurrences. As a construction professional, you must also abide by the RICS ethical standard of 'take responsibility' (RICS, 2020b). This involves acting with care and diligence and raising questions on matters you do not believe to be appropriate. There will be several safeguarding steps that start with simple requests for proof of funds and information on their source. However, if these fail to deliver the answers you require, the next stage may be requests from senior managers and attendance at meetings to discuss the outstanding information required.

While you may never fully discover if money laundering is the intent of the client, if they are unresponsive to the questions and procedures you have in place that are intended to ascertain the sources of any funds, then you will have no choice but to withdraw your organisation's services.

If money laundering is occurring, this could be due to serious criminal intent on behalf the client. It could be to avoid tax, or for the purpose of laundering criminally gained funds from a wide variety of sources. However, regardless of whether you have actual proof of money laundering (if you do, authorities need to be involved as a matter of urgency), if you have suspicions, and these are not alleviated by the answers given from the client, then you will not be able to continue with any working relationships and should look to end all current projects as soon as possible.

8.7 You are self-employed

8.7.1 Scenario 7: Conflict of interests

You are a freelance commercial and project management consultant. You have been hired by a client to advise on and help prepare a procurement framework. Once the procurement framework is complete, your services are no longer used in the comparison of submitted tenders and contract awards.

(?) *Question 11*

You are approached by a main contractor who is preparing a tender for submission to the framework. They have asked you to join their team and help compile the tender they plan to return for entry onto the framework. Given you helped set up the framework, are you able to now work for the main contractor?

Answer 11

This is a complex scenario and one where you will need to proceed with caution. As a freelance consultant you are, in theory, entitled to work for whoever you choose from the offers of employment available. However, this situation is potentially a conflict of interest. The RICS describe conflicts of interest as 'a situation in which the duty of an RICS member … to act in the interests of client … conflicts with a duty owed to another client' (RICS, 2017). In the above scenario, although no longer employed by the client operating the framework, during your role you will have been privy to confidential information and have an awareness of how a commercial advantage could be gained by those companies wishing to submit a tender. If you then use this information for the benefit of the main contractor, this will be operating with knowledge of a conflict of interest and proceeding ahead regardless, for your own financial benefit.

While it may appear unfortunate and against the business 'idea' that as a freelance consultant you can advise any client, it would open the whole procurement process up to additional scrutiny and claims of unfair ethical practice by the main contractor and yourself. Although this would not be illegal and so you could proceed, as discussed throughout this book, ethics go above and beyond what is required by law. The RICS guidance does go further and advises that if you are considering proceeding with an offer that may cause a conflict of interest, then discussions should be had with all relevant parties so they are aware of the potential situation and can provide feedback (ibid.). Elements from a plethora of other professional body codes of conduct also state that members must declare all conflicts of interest. There should also be clear parameters drawn up between yourself and the main contractor about what exactly you are expected to consult on, and on which areas you are unable to offer advice, in order to avoid and minimise any conflict of interest where possible.

It is important to ensure no conflict of interest exists, it is only by not having a conflict of interest in place, either by not accepting a position that comes with a conflict of interest, or by eliminating one that exists, that a professional can meet the RICS ethical standard of 'Acting with integrity', 'Always providing a high standard of service' and 'Act in a way that promotes trust in the profession' (RICS, 2020a).

8.7.2 Scenario 8: Request to provide services

You are a self-employed quantity surveyor with over 20 years' experience working in the construction industry advising clients on how to procure construction projects and services. You have currently completed your latest project and do not have any current projects lined up. You are approached by a new client, a main contractor operating in the infrastructure sector with a lucrative six-month project.

ⓘ *Question 12*

Even though you have no site-based quantity surveying experience, and the experience you do have is in construction and not infrastructure, do you take the role offered?

Answer 12

While the offer may seem perfect in that it is high paying and is available at a time you currently do not have any other work, an ethical professional will need to consider the details of the role's responsibilities and requirements in more detail. For example, you do have extensive construction industry experience, and are a trained quantity surveyor, and as such do have transferable skills. However, you need to be aware of Rule 4 of the RICS Rules of Conduct for Members 'Competence', this rule states that 'Members shall carry out their professional work with due skill, care and diligence and with proper regard for the technical standards expected of them' (RICS, 2020b).

Therefore, as a construction professional, you will need to understand the exact expectations and requirements of the role and be confident in your own ability to successfully meet these requirements. This needs to be balanced with the understanding that no one is an expert at everything, and you may not be expected to know all the 'ins and outs' of all infrastructure elements straight away, plus it may be your transferable skills the company is after as they believe you will be ideal for the role identified. However, you do need to be aware of your own competence and ensure that you are always acting within it, and that your new potential employer is not under the mistaken impression that you are experienced in areas you are not.

8.8 You work for a client

8.8.1 Scenario 9: Withdrawing consent

You are a middle manager for a public sector client organisation. The organisation initially agreed to be part of a collaboration with other industry working partners to create, develop and pilot a new piece of procurement software. However, a new CEO has recently taken over the public sector client and does not want to continue with the collaborative project any longer. They have asked you to stop sharing any further data, retrieve all data shared to date, and ensure a prompt withdrawal from the project.

? *Question 13*

Are you able to follow the CEO's instruction ethically, and stop collaborating immediately, retrieving all data shared to date?

Answer 13

Such issues are always difficult to navigate from a professional point of view, especially if this outcome was never considered prior to the signing of the contractual agreements. Therefore, the first lesson learnt, although it may be too late in this situation to enact it, is to consider all eventualities (no matter how improbable at the time) when drafting the contracts. Therefore, if one party does wish to withdraw consent and cease participation at a future date, there are clear and pre-agreed contractual rules for doing so. In such a situation, you could advise and guide the client through the process.

Professionals should always abide by the RICS ethical standard of 'Provide a high standard of service' (RICS, 2020a) (regardless of their chartered status or professional role), even if they have disagreements with clients or do not fully understand their decisions, especially if such decisions go against any advice they have offered and appear to hinder any progress made in projects. While, as a professional, you can discuss topics, such as withdrawing consent with colleagues in an attempt to fully understand their motivations and advise them accordingly if you think they are mistaken, ultimately if they wish to withdraw consent, this should be actioned as soon as practically possible.

In this instance, while it will be difficult to navigate, you should follow your manager's instruction (if you are satisfied no illegal activity or intentions are involved). You should follow all agreed procedures regarding any data that has been shared, informing the other parties involved in a prompt manner and working together to ensure an efficient and timely withdrawal.

8.8.2 Scenario 10: Behaviour outside of the workplace

You are a senior manager working on behalf of a private sector client. An employee who has recently started and who reports to you has been arrested at the weekend for fighting while intoxicated on a Saturday night. This has not impacted their work as they were let off with a formal police warning on the Sunday morning and then were back in the office Monday as expected. You have only discovered the incident after overhearing some office gossip on Monday morning.

? *Question 14*

Is this an incident you would discuss with the team member involved?

Answer 14

In the first instance, you will need to ensure you abide by the RICS ethical standard of 'treat others with respect' (RICS, 2020a) and not contribute to any gossip or jump to conclusions by prejudging any actions of your staff until you have discussed it with them first. Regardless of any of your personal thoughts at this stage, you must ensure you remain impartial and behave in an ethical manner.

It is advisable to confidentially arrange a meeting with the member of staff, raising the issue of their behaviour outside of work, and double checking all the facts in conjunction with the company procedures and employee terms and conditions of employment. The employee will need to be aware of the expectations of the organisation and of the RICS ethical standards of 'Act in a way that promotes trust in the profession' and 'take responsibility' (ibid.). The latter they will need to embrace when explaining their actions and be accountable for the impacts of their decisions.

Any sanctions issued against the employee should be issued in conjunction with formal procedures that have been highlighted to the employee previously. Some people may argue that any events that happen outside of your professional employment, and not involving any aspect of an employer's business should not be dealt with inside of work, and in this instance as you are unaware of the full details of the incident and no formal police charges were brought, you will need more information before making any decision. However, actions outside of work can result in workplace responses. For example, gross misconduct may be a term inserted in most employment contracts and may cover behaviours outside of the employer's office. A criminal conviction may fall into this category, as may any issues that highlight an individual's profession in a negative manner. During the confidential discussion with the employee in question, company rules should be repeated so they are aware of how external behaviours may be dealt with inside of work, even if they do not directly impact an individual's ability to complete the role required. It is also recommended that more information about any incident is requested as the organisation may be in a position to offer assistance to the employee to help resolve or tackle any issues they are facing proactively, before anything progresses to a more serious stage that may involve more formal professional responses.

8.9 Summary

This chapter has discussed how ethical frameworks can be applied in real-world construction settings and helped construction professionals make better decisions in the face of difficult situations. While there is a difference in theoretical discussions and following the steps outlined here in practice, the very discussion of ethical standard and how to apply them can help illustrate how such standards are not developed in abstract to the construction industry, but have real and practical applications to help professionals deal with a myriad of situations in an ethically correct manner. It is worth noting that sometimes from a range of possible solutions, there is no ideal solution, and no matter the decision made, there will be some degree of harm caused to a stakeholder. In such situations, a utilitarian approach will need to be considered; which decision will lead to the least amount of harm for the least number of people. It is also worth remembering the difference between unethical and illegal acts. While you certainly want to avoid both where possible, sometimes you can only advise, and you may have to be clear in the distinction to clients when offering your advice. If one course of action, although legal is unethical, while another course of action is both illegal and unethical, therefore, in that situation the former is preferred over the latter. It is also worth being aware of ethical standards that exist. If you are a member of a professional body, then you will be governed and held accountable to the ethical standards, and so your awareness and ability to apply them to situations you may encounter will be of paramount importance. It will also be a mandatory requirement to be able to effectively communicate the application of ethical standards if you are applying to be a member of a professional body. Finally, even for those construction professionals who are not members of professional bodies and have no intention of becoming members, the ethical standards of professional bodies will still apply. If a situation ever arose where the conduct of a construction professional was questioned in court, the standards of professional bodies would be used a marker by which the individual's conduct was judged, as professional body standards set the tone for the expected ethical behaviour of all professionals operating in the construction industry.

References

Bibby Financial Services (2019). Subcontracting growth. Available at: www.bibbyfinancialservices.com/about-us/news-and-insights/reports/2019/subcontracting-growth-report-2019

CIPS (Chartered Institute of Procurement and Supply) (2015). Global Standard for Procurement and Supply. Available at: www.cips.org/Documents/Global_Standard/Global_Standard.pdf

ICE (Institution of Civil Engineers) (2004). ICE Code of Professional Conduct. Available at: www.ice.org.uk/ICEDevelopmentWebPortal/media/Documents/About%20Us/ice-code-of-professional-conduct.pdf

IES-C (International Ethics Standards Coalition) (2016). An ethical framework for the global property market. Available at: https://ricstest.files.wordpress.com/2016/12/international-ethics-standards-final.pdf

NBS (National Building Specification) (2015). National Construction Contracts and Law Report. Available at: www.thenbs.com/knowledge/national-construction-contracts-and-law-survey-2015

Prior, G. (2019). Concrete products firms fined £36m for price fixing. Available at: www.constructionenquirer.com/2019/10/23/concrete-products-firms-fined-36m-for-price-fixing/

RICS (Royal Institution of Chartered Surveyors) (2014). Tendering strategies. Available at: www.rics.org/globalassets/rics-website/media/upholding-professional-standards/sector-standards/construction/black-book/tendering-strategies-1st-edition-rics.pdf

RICS (Royal Institution of Chartered Surveyors) (2017). Conflicts of interest. Available at: www.rics.org/globalassets/rics-website/media/upholding-professional-standards/standards-of-conduct/conflicts-of-interest/conflicts_of_interest_global_1st-edition_dec_2017_revisions_pgguidance_2017_rw.pdf

RICS (Royal Institution of Chartered Surveyors) (2019). Countering bribery and corruption, money laundering and terrorist financing. Available at: www.rics.org/globalassets/rics-website/media/upholding-professional-standards/standards-of-conduct/countering-money-laundering-1st-edition-rics.pdf

RICS (Royal Institution of Chartered Surveyors) (2020a). The Global Professional and Ethical Standards. Available at: www.rics.org/globalassets/rics-website/media/upholding-professional-standards/standards-of-conduct/the-global-professional-and-ethical-standards.pdf

RICS (Royal Institution of Chartered Surveyors) (2020b). Rules of conduct for members. Version 7. Available at: www.rics.org/globalassets/rics-website/media/upholding-professional-standards/regulation/rules-of-conduct-for-members_2020.pdf

9 Post-construction ethical dilemmas

9.1 Introduction

Once the building has been completed, the project enters two discrete phases: first, the handover and defects liability period, which can last from 6 months to a number of years if the soft landings protocol has been enacted. Following on from this, the building enters its longest phase: occupancy. During this period the building is likely to undergo phases of refurbishment or extension that would constitute projects as outlined in this book and would once again be subject to all the various phases of professional activity and the subsequent ethical dilemmas the book outlines. Other professionals will now be involved, including building surveying and facilities management colleagues as the building responds to use and begins to deteriorate, thus necessitating maintenance interventions. This of course does not remove the ethical dilemmas; it simply introduces a series of new ethical considerations that this chapter will now examine.

9.2 RIBA Stages 6 and 7

The Royal Institute of British Architects (RIBA) is the leading professional body for architects, established in 1837. In 2020, the latest RIBA Plan of Work was launched, comprising eight work stages. Stage 6 (Handover) and Stage 7 (Use) deal with the occupation of the finished building. During the Handover stage of the project, the building is prepared for occupancy, the contract will have reached practical completion and the client will have taken legal responsibility for the building, but there will still be activity on site, such as commissioning and fitting out of the building. In terms of the main contract, the focus will have moved to resolving latent defects, agreement of final accounts, development of aftercare manuals and potentially a light touch post-occupancy evaluation. As the building moves into Stage 7: Use, the facilities and asset management processes will be implemented and a more detailed post-occupancy evaluation of the building performance will be implemented to verify project outcomes, including sustainability outcomes (RIBA, 2020).

9.3 Professional ethics

The following are a series of real-life examples which pose ethical dilemmas that you may face during your professional career. Although these are focused on the building during its occupancy phase and will subsequently only have limited relevance to the construction project manager and quantity surveyor, they will resonate with other branches of the surveying profession, especially those who will see their role spanning the Construction and Infrastructure and Property professional groups of the RICS, including, but not limited to, the general practice surveyor, the building surveyor, the facilities manager and the asset manager. As with earlier chapters, you may consider some of the examples presented in this chapter to be very minor in impact but whether they are major or minor, all breaches of professional ethics erode the public's perception of our sector and are thus treated equally seriously by the professional bodies. As with previous chapters, this chapter takes these frameworks and applies them to a set of scenarios to help navigate the ethical dilemmas you may face or be asked to advise upon throughout your professional career.

9.4 Professional ethics post-construction

9.4.1 Scenario 1: Discovery of latent defect

You are currently employed as a construction manager, working on a multi-phased project constructing luxury apartments for both the private rental and speculative sales market. Phase one of the project concluded six months ago and has now been fully sold out by the developer. Phase two has recently achieved completion, however, during the pre-purchase inspection of one of the apartments, a customer has noticed a small amount of water ingress above the French doors that lead out onto a balcony which has been finished with external tiling. Concerned about the long-term implications of the defect in terms of water ingress from the balcony above, the purchaser has requested the defect be addressed before exchanging contracts on the purchase.

? *Question 1*

The developer has approached you as site manager to ask if you can investigate the issue. However, the subcontractor responsible has refused to proceed without payment for the investigation, arguing there is no defect. How do you proceed?

Answer 1

It is clear in this case the subcontractor is appearing to argue the difference between a patent defect and a latent defect. This is significant, as the project has achieved practical completion and has entered the defects liability period. Thus, the project has been free of patent defects with the developer inspecting the work and not identifying the error. Clearly if this issue had arisen prior to practical completion, the subcontractor could have been instructed to open up the work and, depending on what is discovered, would have been paid for the work. However, it is clear the waterproofing from the balcony has been covered with the tiled finish, so it would fall within the remit of a latent defect and contractually the subcontractor can be required to investigate it.

⑦ *Question 2*

Having undertaken the inspection, you have identified a defect in the waterproofing detail which has caused the water ingress into the building. The subcontractor did not take on the design responsibility for the balcony finish. Your commercial director has agreed with the sales team that for those properties in Phase 3, remedial work will be undertaken to correct the defect. When you asked (privately) about rectification works to Phases 1 and 2, your commercial director instructed you to 'keep your mouth shut, there is not enough profit in this project to go looking for work'. You are worried about this approach, what should you do?

Answer 2

In this situation, a clear design defect has been identified. It is obvious this defect will have been present in all phases of the project, given the works are Design and Build, the responsibility for this defect rests with the constructor, who in this case is also the developer. It is not uncommon in construction for constructors to attempt to hide defects and only respond when someone challenges them legally under the Defective Premises Act 1972 or the Consumer Rights Act 2015. In terms of the latter, Power (2015) advises the house building sector that CRA legislation effectively

> renders ineffective any contractual attempts to exclude or restrict liability to consumers for breach of implied terms as to fitness for purpose and satisfactory quality of goods … and it provides additional and extended remedies (including rights to reject, repair, replacement and price reductions/refund) where businesses fall short.

This very clearly confirms the legal position that the developer is liable for this latent defect and has an obligation to address it, but this also appears to put the responsibility for identification on the consumer.

It is, therefore, clear the developer is hoping the latent defect will not appear, or not appear until responsibility transfers to the third-party insurance company who warranted the building. Ethically, however, neither positions are ethically acceptable, given the duty of care and duty of professionalism owed to the purchasers of the development. Rule 5.3 of the CIOB's Rules and Regulations of Professional Competence and Conduct obligates members to 'ensure, when undertaking any other construction related activity, that all such work is in accordance with good practice and current standards and complies with all statutory and contractual requirements' (CIOB, 1993). In this example, it is clear, having discovered the defect that the work does not comply with good practice, current standards or the implied obligations owed to the purchasers of the apartments, thus constituting a clear breach of the CIOB's ethical framework. RICS opines a similar position with Rule 3 of the Rules of Conduct for Members asserting that members 'shall at all times act with integrity and avoid conflicts of interest and avoid any actions or situations that are inconsistent with their professional obligations' (RICS, 2020). So, it is clear there is a breach of ethics here, in terms of both the integrity of the professional and also in terms of the duty of care afforded to all stakeholders.

Additionally, the organisation will also have professional obligations, in this instance, as the development is being sold directly to the public, the developer will undoubtably be offering a home warranty (as this is essential for customers' mortgages) via an organisation such as NHBC, Premier Guarantee, LABC Warranty or Checkmate. As a condition of offering these warranties, they will be a registered housebuilder and consequently affiliated to the Consumer Code for Home Builders, which places a number of ethical obligations on the organisation, including explaining that the housebuilder is 'responsible for remedying relevant defects arising under the Home Warranty two-year defect period'. However, the code does not specifically obligate the organisation to inform previous customers that they are aware of latent defects.

However, in this situation, it would be very difficult for you to approach the customer and freely declare the issues, as this would likely put you in breach of your contract of employment and other duties you owe to your employer. It is regrettably clear the obligations of the organisation necessitate the defect to be first discovered before further action can be take. However, in this situation, the RICS (n.d.b) advises that you use your best endeavours to 'encourage the firm or organisation you work for to put the fair and respectful treatment of clients at the centre of its business culture'. The RICS also advise that

> If you think something is not right, [be] prepared to question it and raise the matter as appropriate with your colleagues, within your firm or the organisation that you work for, with RICS, or with any other appropriate person, body or organisation.
>
> (ibid.)

Regrettably, at this stage, there is little more you can do, if the organisation is determined to behave unethically, I am afraid there is little you will be able to do to change that.

? *Question 3*

How would you address this situation if you were the employer's agent and you did not identify the defect?

Answer 3

In this situation, the ethical obligations are exactly the same as those outlined in question 2 in terms of personal professional obligations. However, unlike in question 2, you do have some ability to take action in this situation. First, it is important to resolve the issue of liability, are you responsible for missing the defect? To answer this, the RICS guidance note 'Employers Agent: Design and Build' addresses patent and latent defects, advising first that not identifying a defect does not in itself constitute a breach of your professional duties and obligations. The guidance advises:

> The employer's agent will only list what are usually defined as patent defects, i.e. those that are observable from reasonable inspection at the time, examples being a defective concrete finish or an incorrect paint colour, and will not include what are usually defined as latent defects, which may be hidden from reasonable inspection and may come to light at a much later date, examples being some structural defects.
>
> (RICS, 2017a)

As the defect in this instance relates to the waterproofing substrate that would have been covered by the balcony tiling, it would have been impossible to identify during inspections of the work at handover. So now you have become aware of the defect, what action can you take? Well, the same guidance note advises that liability and rectification of these defects will be resolved legally. However, the employer's agent will return the duty of care to the employer, and as such you should notify the employer of the presence of the defect. If your services have been retained for later phases of the project, you should also 'assist the employer in making informed decisions on the potential defect' (ibid.). Alternatively, if your employment has ended, you could also be 'called on to provide information pertaining to any claim against the contractor relating to a latent defect after the employer's agent's appointment has ended' (ibid.).

9.4.2 Scenario 2: Post-occupancy evaluation

Employed as an employer's agent for a Design and Build project, your client has asked for your opinion on the use of a post-occupancy evaluation for their latest project. You are aware of the massive benefits a post-occupancy evaluation can provide in terms of feedback and future refinement of projects. However, you are also acutely aware that they are expensive to conduct and do not attract a fee income for the practice. Your client is inexperienced, so you could advise them either way.

ⓘ *Question 4*

How would you advise your client?

Answer 4

Hey et al. (2017) have identified that professionally there is minimal uptake of the post-occupancy evaluation (POE), with only 9 per cent of architectural practices advising that they undertook them for their clients. Often this was the result of a myriad of barriers including financial (lack of fee income for the work) but also due to confusion about what constitutes a POE survey, the lack of incentivisation for improvement in the current focus on cost implemented by employers and concerns about the liabilities the practice may face (ibid.). The argument against adoption could be presented on this basis of general custom and practice.

Even so, the employer in this situation has expressly asked for a professional opinion on the value of the survey to their business and the benefits this could bring to future projects. The benefits associated with implementation of POE are clear, writing for the RIBA, MacDonald and Forth (2020) have highlighted the major benefits that adoption of POE would bring, making it easier to learn from both successful and more troubled projects in a way that would enhance the quality of the built environment. In the same report, MacDonald and Forth advocate that, in terms of cost, a POE would add between 0.1 and 0.25 per cent to the overall professional fees on projects where it is adopted. However, they also recommend clients include this in their 'contractual obligations with the project team from the outset' (ibid.), thus ensuring this is costed appropriately into the services provided. So, in this case, it is clear the work is not in the scope of the original contract and would thus constitute a change and attract an additional fee should the client wish to proceed.

The two sides of this particular issue have been highlighted above, in the first instance the lack of adoption would seem to make a custom and practice argument for advising against adoption. However, the RICS guidance note 'Employer's Agent: Design and Build' (RICS, 2017a) makes it clear that the surveyor should always 'provide advice based on his or her knowledge and experience to assist the employer in making informed decisions'. This perspective aligns with the more general ethical duty the surveyor owes to 'act with integrity' and provide the client with a high standard of service. Thus, in this situation, regardless of the wider industry practice associated with adoption, the RIBA have made explicit the benefits clients can obtain from undertaking a POE, so the advice given in this situation should be to proceed with the survey, but be aware there is likely to be an additional fee for this service.

⑦ *Question 5*

Given the level of trust you have developed with the client; they have asked if you would be willing to undertake the POE. Your practice is keen to make the additional fee and is pressuring you to accept. However, you have never completed a POE before, would you take on this work?

Answer 5

This is not an uncommon situation and clearly provides an easy approach for the client, they trust you, given your relationship and they are not likely to understand the specifics of how a POE is executed. The challenge in this situation is that this would provide additional fee income and would cement the relationship with the client. However, as an ethical professional, you need to consider whether you have the competence to complete this task. Rule 4 of the RICS Rules of Conduct for Members addresses 'Competence'; this rule states: 'Members shall carry out their professional work with due skill, care and diligence and with proper regard for the technical standards expected of them' (RICS, 2020). In a similar vein, Rule 7 of the CIOB Rules and Regulations of Professional Competence and Conduct makes clear that 'members who are not competent to undertake part or all of a particular advisory service shall either decline to give advice or secure appropriate competent assistance' (CIOB, 1993). Yet in both situations, the rules also make clear that it is for the professional to decide on their own levels of competence. For example, if you are a quantity surveyor looking to provide project management services, the RICS position makes clear that you do not need to be re-assessed via the APC for this role, but you need to evidence your competency, say, by completing a post-graduate qualification and working alongside an experienced project manager for a reasonable period of time to gain project management experience. However, in this situation, you have no experience of undertaking post-occupancy evaluations. Therefore, you are unlikely to fully understand the exact expectations and requirements of the role and be confident in your own ability to successfully meet these requirements. In this situation, you, therefore, need to be aware of your own competence and ensure that you are always acting within it, so the sensible course of action would be to explain to the client this is outside your sphere of competence but you can arrange for another professional to undertake the service.

9.4.3 Scenario 3: Contradictory survey reports on a property

Employed as a surveyor holding AssocRICS membership, including registered valuer status, you have been tasked with undertaking a mortgage valuation on a residential property for a lender, with whom your organisation has an established panel relationship. You visit the property at the allocated appointment and after asking the vendor a series of questions you proceed to inspect the property and take the photos required by the mortgage company's valuation framework. During the inspection it becomes clear that the property, although currently mortgaged, is of a non-standard construction and thus is not mortgageable. You return the valuation to the mortgage company, advising the property has no mortgageable value due to the nature of its construction. Some days later, your manager calls you into the office to say the practice valued the house three years ago as mortgageable and the vendor is now threatening to sue the practice for professional negligence.

? *Question 6*

At the meeting, your director questioned you on why you deemed it to be non-standard, suggesting the previous surveyor, who was FRICS, was a lot more experienced and thus would not have made such a catastrophic error and maybe you have simply misinterpreted the building through your lack of experience. He then asked you to admit to the error and retract the valuation. How would you respond?

Discussion Point 9.1

Before the position of the surveyor is reviewed, do you feel the director's handling of this situation is appropriate? Has that director breached his professional and ethical responsibilities? We will look at this in question 7 below, so please make some notes and see how your thoughts compare.

Answer 6

This question opens up three different strands of answer, the ethics and professionalism of (1) the second surveyor who undertook the survey; (2) the ethics and professionalism of the original surveyor; and (3) finally, the ethics of the director. These positions will be explored in sequence.

The second surveyor

In response to this pressure you first need to establish whether you have made an error. Page 1 of the practice note 'RICS Building Survey' addresses the issue of negligence, directly advising that 'when an allegation of professional negligence is made against a surveyor, the court is likely to take account of any relevant practice notes in deciding whether or not the surveyor acted with reasonable confidence' (RICS, n.d.a). You should therefore return to the notes you made during the mortgage valuation survey and the photographs you took to once again review the nature of the building's construction. During the site inspection, the guidance note to 'Surveys of Residential Property' makes it clear that if you suspect something is out of the ordinary or non-conventional, the surveyor 'must take reasonable steps to follow the "trail"'. These 'reasonable steps may include extending the extent of the inspection and/or recommending further investigation' (RICS, 2016: 11). So, it would be important to ensure you did seek a resolution to your suspicions during your physical inspection and this is expected to be recorded in your inspection notes.

Additionally, in the comments made by your director, your competence has been indirectly called into question. At the start of the 'Surveys of Residential Property' guidance note, the authors advise that in order to

> provide a satisfactory service at all levels, the surveyor must be qualified (AssocRICS, MRICS or FRICS), competent and have:
>
> - sufficient knowledge of the tasks to be undertaken and the risks involved
> - the experience and ability to carry out their duties in relation to the appropriate level of service;
> - the ability to recognise their limitations and take appropriate action where this is found to be inadequate.(ibid.: 1)

The guidance note also makes it clear that 'the surveyor needs to be familiar with the nature and complexity of the property type and the region in which it is situated' and that this would include 'knowledge of common and uncommon vernacular housing styles, materials and construction techniques'. Additionally, Rule 4 of the RICS Rules of Conduct for Members addresses 'Competence'. This rule states that 'Members shall carry out their professional work with due skill, care and diligence and with proper regard for the technical standards expected of them' (RICS, 2020). So before departing to survey the property, it would be expected the surveyor would have deemed themselves to be sufficiently competent to undertake the survey, regardless of their RICS membership class, this is a specific requirement of Rule 4 as noted above. Therefore, in executing the survey, it would be expected the surveyor has operated within their clear sphere of competence. To be pressured to change this opinion is not acceptable.

Discussion Point 9.2

Put yourself in the position of the FRICS qualified senior surveyor, who may well have made an error on this occasion when originally surveying the property. What do you do?

The original surveyor

At this stage in this scenario, it must be noted that the origin of the error has not been determined but let us assume the original surveyor made an error in the initial survey. What should that person do? Remember at this point, they are a senior surveyor, they have likely trained a number of junior colleagues and this error is going to be extremely embarrassing, not to mention the legal action that is going to follow likely from the lender and also the client. Finally, the RICS may well also review the situation. What should they do?

In this regard, the RICS position on ethical conduct is extremely clear. The RICS Global Professional and Ethical Standards identify the obligation to 'take responsibility' under this ethical obligation, it is expected members will be accountable for all their actions, they won't blame others if things go wrong, and if you suspect something isn't right, be prepared to do something (RICS, n.d.b). Furthermore, this also requires the surveyor to respond in 'an appropriate and professional manner [to complaints] and aim to resolve the matter to the satisfaction of the complainant as far as you can'. So, in this instance, assuming the first surveyor made the error, they need to take responsibility for that error and the practice also needs to take responsibility for this and notify their professional indemnity insurer of the likely claim. Should the lender or property owner make a formal complaint to the RICS, Rule 9 of the RICS Rules of Conduct for Members requires that the surveyor would 'cooperate fully with RICS staff and any person appointed by the Standards and Regulation Board' (RICS, 2020) and accept any outcome that is deemed applicable. However, if the surveyor has acted professionally, followed the RICS guidance notes and not acted beyond their own competence, they have nothing to fear from a genuine error of judgement. The professional indemnity insurer and the courts will address the financial damage caused by the error. The individual surveyor, however, may well feel further professional development is an appropriate measure to ensure the error does not reoccur. This would be the same position, should the junior surveyor have made a mistake with the second survey, although the practice may well take additional steps, and these will be outlined below.

The practice director

The final stage in answer 6 focuses on the director of the practice, who has received the complaint from the customer, who has threatened to sue the practice for professional negligence. At this point, this is merely a factual statement. However, at the meeting, the director has also opined that the second survey is erroneous and, given the vendor is now threatening to sue the practice for professional negligence, they want you to admit to making an error and bring your survey into line with the first surveyor's view, suggesting the surveyor had 'simply misinterpreted the building through a lack of experience'. This position is very clearly a complete breach of professional ethics.

Taking the various issues in turn, they give two different perspectives on the situation. The first perspective is that the firm has knowingly or unwittingly asked an inexperienced surveyor to complete a survey outside their scope of competence. While the personal ethics of this have been addressed earlier, as the surveyor, if lacking experience, has breached their professional duty of care and acted unethically. It is also clear the practice also needs to take responsibility. Rule 6 of the RICS Rules of Conduct for firms requires firms to 'have in place the necessary procedures to ensure that all its staff are properly trained and competent to do their work' (RICS, 2020). If they are suggesting the surveyor did not have the required experience, the firm should have been aware of this, and they should not have allocated the survey to them to complete. The firm will also, at this point have breached Rule 5 relating to the 'services' they offer, which requires that the firm carries out 'its professional work with due skill, care and diligence and with proper regard for the technical standards expected of it'. Once again, this requirement will have also been breached. Consequently, the firm will have to rectify the situation and notify their PI insurer of the possibility of a claim.

The other perspective on this is that the manager is seeking to implement some form of cover-up, and to remove the possible claim by issuing a knowingly false survey that supports the position adopted in the previous report. This would leave the practice open to various legal interventions and possible criminal sanctions. Such a decision would also have very clear ethical breaches. Looking first at the RICS Rules of Conduct for firms, Rule 3, 'professional behaviour' requires the firm to 'at all times act with integrity and avoid conflicts of interest and avoid any actions or situations that are inconsistent with its professional obligations' (ibid.). Any attempt to cover up the earlier error would clearly constitute a breach of professional conduct. The firm would also find itself in breach of Rule 4 'competence' which obligates the firm to 'carry out its professional work with due skill, care and diligence and with proper regard for the technical standards expected of it' (ibid.) and, finally, Rule 5 related to 'service', which requires the firm to 'carry out its professional work with expedition and with proper regard for standards of service and customer care expected of it' (ibid.). Any form of cover-up to protect the firm would be a major breach of the firm's duty of care to the customer and its professional obligations. In addition to the code of conduct for firms, there are also clear breaches of the RICS Global Professional and Ethical Standards, including a failure to act with integrity, and failure to take responsibility for the error the firm possibly made in the first survey. And failure to provide a high standard of service to the lender, vendor and purchaser of the property. The failure to enhance trust in the profession, as once the actions became known, would damage the general public's trust in the profession. Finally, there is also the failure to treat a junior colleague and others with respect, by asking them to admit to and take the blame for a potentially non-existent error, this hardly demonstrates fair and respectful treatment.

Discussion Point 9.3

Put yourself into the position of the director of the firm, what should they have done?

In answering this final point, the RICS Rules of Conduct for firms make it clear that all regulated firms are required to 'operate a complaints-handling procedure and maintain a complaints log' (RICS, 2020), so upon receipt of the complaint, the director should investigate, asking both surveyors why they feel the property was or was not of a standard construction and potentially ask a third surveyor to form an independent objective opinion of the construction. This would then determine where the error lay. The director could then determine the action to be taken, observing the ethics or otherwise of the original surveyor's practice. Finally, the director should notify their professional indemnity insurers of the potential claim and follow the guidance they provide in seeking a resolution.

9.4.4 Scenario 4: Term contracts

You have recently been employed as a freelance quantity surveyor to review the administration of a term contract that a large brewery has with a maintenance company, addressing small works projects in its estate of pubs located through the UK. The company is concerned with the volume of claims being submitted for works to some of its establishments which seem to be completely out of sync to those experienced with the previous provider. Having raised the issue with the contractor, they have robustly defended their claims, suggesting they are only claiming for work completed. However, to resolve the dispute, the brewer has appointed you to administer the contract and to monitor and approve expenditure and ensure this is reconciled with the maintenance instructions from the central customer services team.

Ⓡ *Question 7*

In the second month of this process, the contractor has presented their claim with quite a lot of work detailed, however, they cannot present work orders from the brewer's customer service team for all the work they have undertaken. When asked, they have said landlords are asking their maintenance operatives to complete the work, suggesting it is urgent and they have reported it to the brewery. How would you proceed?

Answer 7

In this situation, you may well feel that paying for the additional work is correct, work that has been completed by the firm and has been required, as the pub landlord/tenant has identified the work as urgent to the maintenance operative. Equally, you may well think that the provisions of the contract should be adhered to and payment should only be made for authorised work, with a clear governance process now in place within the client organisation to allow landlords to report maintenance needs and for these to be authorised and then instructed and completed. If maintenance operatives are making these decisions unilaterally, this is not acceptable as it removes all control and monitoring from the brewery who need to ensure that only works for which they are liable are addressed.

However, from an ethical perspective, which of the two positions is correct? Is the robust execution of the contract the correct approach? If we consider the first of the five RICS Global Professional and Ethical Standards (RICS, n.d.b) 'Act with integrity', can we describe a construction industry professional who purposefully does not pay another company for work that has been carried out as acting with integrity? There is often a misunderstanding when it comes to the administration of contracts, especially when custom and practice have in effect deemed this process acceptable. In this instance, however, would it be acting with integrity to consider the purpose of the instruction, as you have been employed to take control of this contract and stem the overspend? Evidently, maintenance operatives undertaking any work they feel they can do and then charging the client sits at the core of the issue you have been employed to address. The question is, does retrospective rejection of the claim meet the ethical responsibilities you hold? Rule 3 of the RICS Rules of Conduct for Members requires members to 'at all times act with integrity and avoid conflicts of interest and avoid any actions or situations that are inconsistent with their professional obligations' (RICS, 2020). In this situation, those obligations are owed to the brewery, in which case, refusal of the claim would be the correct course of action.

However, the issue of custom and practice must also be addressed. The contract does allow 'wriggle room' for the surveyor to make the payment, therefore, it would make sense to establish how this change in process has been communicated and when. Has the client informed the contractor of the new contractual process for payment and given a date for this to commence, with enough notice for this to take effect? Have they also notified their landlord tenants of the new process for requesting and auditing repairs? If the client only did this, say, 7 days go, then can the contractor be expected to apply this retrospectively? This would clearly not be a fair and reasonable request and thus not allowing the payment for work completed prior to this notification would not be fair to the maintenance company. Imposing this to benefit the brewery would not constitute 'acting with integrity'.

On the other hand, if the brewery issued the new procedures two or three months ago and advised they would take effect after the next monthly valuation and payment for completed works, while also simultaneously notifying all landlords that maintenance teams will only complete pre-authorised work and this is to be the procedure for registering maintenance requirements from now on, then it would seem fair and reasonable to act in the client's best interests and stem the abuse of practice that has become evident and thus refuse to pay for the unauthorised work. In this situation, withholding payment would be the correct course of action as allowing the maintenance contractor to maximise turnover at the expense of the client is not acting with integrity.

9.4.5 Scenario 5: Receipt of gifts

You are a clerk of works working for a medium-sized social housing organisation and your role is predominantly focused on inspecting major planned maintenance projects and new housing developments. As part of your role, you regularly meet with the main contractor's site teams both during and after works have completed. On one such job, you developed a very good working relationship with the construction manager and noted the high quality of workmanship on the project which led to little or no snagging at practical completion.

? *Question 8*

The director of the company wished to personally thank you for helping to deliver a very successful project and presented you with a new phone worth circa £1,200. Do you accept or decline the gift?

Answer 8

Regardless of the size or value of the gift, the RICS say a professional must not 'accept directly or indirectly, anything that could constitute a bribe' (RICS, 2019) with a bribe described by the same document as 'the offer, promise, giving, demanding or acceptance of an advantage as an inducement for an action that is illegal, unethical or a breach of trust'. In this instance, the offer of a mobile phone does not initially fulfil the definition above.

Before you proceed, however, questions must be asked of the timing and cost of any gift. Most companies will have established practices for this, and these will consider the legal implications of any potential bribe. For instance, a lot of firms will apply a monetary value cap of, say, £15 and also operate a central recording system for all such gifts. Clearly an expensive gift such as a £1,200 phone should always, as a matter of course, be immediately declined

? *Question 9*

The director of the company, responds to your polite decline of his gift angrily, saying that it is a personal insult and an insult to his culture where it is customary to exchange gifts on the completion of a successful relationship. Should you just give in and accept the gift?

Answer 9

In this situation, you are likely to be worried about the implications of cultural insensitivity and fear refusing the gift could be construed negatively. You are also worried that this could somehow breach the RICS ethical standards, as you are failing to respect the director's culture and as such you may well not be 'treating others with respect'. This makes the decision a little more difficult, but the position adopted in response to question 8 remains the correct position. Ethically, the RICS Global Professional and Ethical Standards (RICS, n.d.b) make it clear that acting with integrity means 'not offering or accepting gifts, hospitality or services, which might suggest an improper obligation'. If you were to explain this, along with the wider implications of accepting the gift to the director in a courteous and professional way, this would ensure you continued to act with integrity while also ensuring the obligation to treating others with respect, in this case, the director, was met.

9.4.6 Scenario 6: Conflict of interest

After qualifying as a chartered construction manager six months ago, you were offered a new post as a senior building surveyor working for a maintenance contractor. Having taken on this role you successfully improved some of the company processes and are well regarded by your colleagues and senior management. A couple of weeks ago, you discovered your new firm has been asked to tender for a major repair and maintenance partnership contract by your previous employer. Sensing a competitive advantage, your new employer has now asked you to give them some insider tips, suggesting you need to reflect on where your loyalty lies.

? *Question 10*

How should you respond? Should you disclose potentially commercially sensitive information, or should you respectfully decline?

Answer 10

This is a very difficult situation to find yourself in, on one hand, you have been at the firm for 6 months and you have settled into your new role well. You departed from your previous employer on good terms, but you did not sign any form of non-disclosure agreement. In reality, this scenario has uncovered divided loyalties coupled with a potential conflict of interest.

In theory, you are entitled to change employers and also your previous knowledge cannot be forgotten, equally it is possible a potential conflict of interest could exist. The RICS describe conflicts of interest as 'a situation in which the duty of an RICS member … to act in the interests of client … conflicts with a duty owed to another client' (RICS, 2017b). In the above scenario, although no longer employed by the client, during your role you will have been privy to confidential information and have an awareness of how a commercial advantage could be gained by those companies wishing to submit a tender. If you then use this information for the benefit of your current employer, this would mean you are operating with knowledge of a conflict of interest and proceeding ahead regardless, for your own benefit.

In this situation, while the outcome may well go against the principle that you can work for whomever you wish, should any party discover that you have disclosed confidential information, this could open up the tender process to additional legal scrutiny or may lead the client to reconsider the position of your company, both on this contract and any future contracts. Although this would not constitute a legal breach, there are clear ethical concerns, Rule 5.3 of the CIOB's Rules and Regulations of Professional Competence and Conduct (CIOB, 1993) state that members must 'not divulge to any person, firm or company any information of a confidential nature relating to the business activities or processes of their employer or client acquired during the course of their work'. As a result, you should respectively decline the request, advising your boss that you cannot disclose the information requested as this would constitute a clear breach of your professional obligations.

9.4.7 Scenario 7: Final account

Works have recently completed on a large distribution warehouse procured through Design and Build and let on the Joint Contracts Tribunal (JCT) D&B 2011 standard form of contract. During the course of the project, the initial contractor appointed to the project ceased trading and entered into insolvency due to major cash flow challenges. A step-in right written into collateral warranties and other safeguard mechanisms allowed you to retain the architects for the project and novate these to the new contractor, appointed after a competitive tender process to conclude the project. However, now the project has reached practical completion, you have been tasked with agreeing the final account.

? Question 11

Knowing the final account will be complex, the client has asked you to ensure you recover their costs in terms of appointing the new contractor, but they have also asked you to adjust the final account so that a couple of additional items of work that occurred but are not documented too well are not included to keep the project costs down. How should you proceed?

Answer 11

Insolvency of contractors is a rather common feature in the construction sector. Analysis of company insolvency by Creditsafe shows the sector has one of the highest levels of company failure and insolvency in the economy, with records showing 2,123 insolvencies in the construction sector between January and September 2020, which represents 14.56 per cent of all company insolvencies (Creditsafe, 2020). As such, it is likely all surveyors will encounter insolvency at some point in their career.

In this instance, the contractual requirements have been addressed and the project re-let, however, there is now the need to agree final accounts. The

RICS guidance note 'Termination of Contract, Corporate Recovery and Insolvency' advises that at the time of the insolvency the surveyor would have issued the insolvent contractor with a 'valuation statement with the current position of the works, detailing all risks, liabilities and claims' (RICS, 2013: 16). It is also expected that the surveyor will have also 'provided information on the position of the project/works in respect of finance, progress and management' (ibid.: 20) to the insolvency practitioner appointed to manage the firm's affairs and close down the business. At this stage in the project, the surveyor will now follow up these initial statements with two final account documents, the first is a 'notional final account'. The RICS guidance note advises that

> As a result of the insolvency, the notional final account will be higher than the anticipated final account before insolvency. This is because the notional final account will recognise the cost incurred by the party at the time of insolvency and will build on that cost to show the cost to complete (had the original contractor completed its works). The difference between this amount and the original sum as a commitment to complete the works is a debt, the same amount, or a gain. Generally, the difference is a debt since appointing a new party to complete the construction project has inherent additional costs.
>
> (ibid.: 19)

This notional account will then be followed with a second financial report, the 'Completion final account' that will include the full effects of 'any claims, such as loss and/or expense, disruption, prolongation, acceleration, etc.' (ibid.: 19). Finally, the RICS guidance note advises that, given the agreement of the completion final account settles a number of contractual matters and can be 'construed as compliance with the contract', it is important this document is as 'complete and as fully substantiated as possible' (ibid.: 20).

Returning to the question, in this instance, the client has asked you to adjust the final account so that a couple of additional items of work that occurred but are not documented too well are not included to keep the project costs down. The guidance above from the RICS outlines the robustness and importance of this process and its accuracy. However, the contract also introduces some 'wriggle room' as the contract allows for verbal instructions to be used under clause 4.3.2 (RICS, 2015). So, there is scope to address the missing instructions, but this responsibility rests with the contract administrator, not the quantity surveyor. The RICS guidance note 'Final Account Procedures' makes it clear authorising undocumented change is not the role of the quantity surveyor, your role in the final account is to simply 'value variations in accordance with the contract and adjust the final account as necessary' (ibid.: 4).

If, however, you had also taken on the role of contract administrator, then you need to ensure you reach an ethical and impartial decision on this issue. While the contract provides the manoeuvring space to either disallow the additional costs due to the lack of a formal instruction or to resolve this with the contractor at this stage, using what the RICS guidance note refers to as a 'wrap-up' instruction to formalise these few outstanding items (ibid.: 12), the decision rests with you.

Discussion Point 9.4

Regardless of the contractor's insolvency, would you consider it fair and reasonable to withhold payment for works completed in good faith? Would this meet your legal and ethical obligations of both acting impartially and but also acting with integrity?

It could hardly be seen as being impartial to follow the employer's instruction and deliberately withhold payment for completed work. Furthermore, adopting such a position, to favour the employer during the final account process would very clearly breach Rules 3, 4 and 5 of the RICS Code of Conduct for Members. Rule 3 obligates members to 'at all times act with integrity and avoid conflicts of interest and avoid any actions or situations that are inconsistent with their professional obligations'. Rule 4 obligates members to 'carry out their professional work with due skill, care and diligence and with proper regard for the technical standards expected of them'. Rule 5 finally obligates members to 'carry out their professional work in a timely manner and with proper regard for standards of service and customer care expected of them'. In this situation, the guidance note 'Final Account Procedures' makes it very clear that these issue must be resolved and that some tidying of the instructions at the end of the project may well be needed to capture outstanding verbal instructions, that are contractually permitted. As such, you should disregard the pressure from the employer and ensure all works are fairly captured in the final account and that all works verbally instructed are supported by a wrap-up written instruction to allow the quantity surveyor to include this expenditure in the final account.

9.5 Summary

This chapter has explored how ethical frameworks can be applied to real-world construction problems that could occur in the final stages of a project as it moves from handover just after practical completion into occupation, while also helping you and your fellow construction professionals make better decisions in the face of difficult situations. As other chapters have espoused, there is without a doubt a difference between the presentation of hypothetical ethical situations that are responded to in a theoretical discussion and real life, with the benefits of hindsight, time and research to reach correct responses. However, these situations are drawn from the real experiences of construction professionals, they attempt to articulate how ethics can influence your professional life. The authors acknowledge that the realities of those experiences in professional life, with the pressure of time applied to the decision, will often be different. Not because construction professionals are behaving unethically, rather the pressure of time and the immediacy of response can prevent them from providing comprehensive responses. It is hoped this review of some of the myriad of ethical issues that arise in this aspect of the construction project will at least assist you, the construction professional in understanding how these issues can be addressed competently and ethically. It is hoped these examples will also help those seeking chartership who, when in a high-pressure situation, are presented with questions of ethical judgement that, in the case of the RICS, can mean the difference between a successful and unsuccessful outcome, so they will be in a stronger position to respond. The key take-away message from this chapter is ethical dilemmas are challenging, complex and often have no single clean answer. It is our duty as professionals to have a thorough understanding of professional ethics and to be able to apply these to the situation we face. Ultimately, as we do in our wider life, you will develop an intuition for the correct ethical decision and will implement this, often without giving it a second thought. Regrettably, however, not all professionals adhere to these ethical norms of behaviour and the ultimate test becomes one of legal responsibility under the tort of negligence or the view of the panel in an employment or professional body hearing. It is, however, our hope that the readers of this text never find themselves in that predicament.

References

CIB (Chartered Institute of Building) (1993). Rules and Regulations of Professional Competence and Conduct. Available at: www.ciob.org/sites/default/files/Rules%20&%20Regulations_0.pdf

Creditsafe (2020). UK business insolvencies. Available at: www.creditsafe.com/gb/en/blog/reports/insolvencies.html

Hey, R., Samuel, F., Watson, K.J. and Bradbury, S. (2017). Post occupancy evaluation in architecture: Experiences and perspectives from UK practice. *Building Research and Innovation*, 46(6).

MacDonald, P. and Forth, A. (2020). *Post Occupancy Evaluation: An Essential Tool to Improve the Built Environment*. London: Royal Institute of British Architects.

Power, L. (2015). Consumer rights focus for housebuilders and developers. Available at: www.walkermorris.co.uk/publications/consumer-rights-focus-for-housebuilders-and-developers/

RICS (Royal Institution of Chartered Surveyors) (2013). Termination of contract, corporate recovery and insolvency. Coventry: Royal Institution of Chartered Surveyors.

RICS (Royal Institution of Chartered Surveyors) (2015). Final account procedures. Coventry: Royal Institution of Chartered Surveyors.

RICS (Royal Institution of Chartered Surveyors) (2016). Surveys of residential property, 3rd edn. Coventry: Royal Institution of Chartered Surveyors.

RICS (Royal Institution of Chartered Surveyors) (2017a). Employer's agent: Design and build. Coventry: Royal Institution of Chartered Surveyors.

RICS (Royal Institution of Chartered Surveyors) (2017b). Conflicts of interest. Available at: www.rics.org/globalassets/rics-website/media/upholding-professional-standards/standards-of-conduct/conflicts-of-interest/conflicts_of_interest_global_1st-edition_dec_2017_revisions_pgguidance_2017_rw.pdf

RICS (Royal Institution of Chartered Surveyors) (2019). Countering bribery and corruption, money laundering and terrorist financing. Available at: www.rics.org/globalassets/rics-website/media/upholding-professional-standards/standards-of-conduct/countering-money-laundering-1st-edition-rics.pdf

RICS (Royal Institution of Chartered Surveyors) (2020). Rules of Conduct for Members, Version 7 with effect from 2nd March 2020. Available at: www.rics.org/globalassets/rics-website/media/upholding-professional-standards/regulation/rules-of-conduct-for-members_2020.pdf.

RICS (Royal Institution of Chartered Surveyors) (n.d.a). RICS building survey practice note. Coventry: Royal Institution of Chartered Surveyors.

RICS (Royal Institution of Chartered Surveyors) (n.d.b). The Global Professional and Ethical Standards. Available at: www.rics.org/globalassets/rics-website/media/upholding-professional-standards/standards-of-conduct/the-global-professional-and-ethical-standards.pdf.

10 Conclusion

Ethics can be a diverse, subjective and personal concept. In the construction industry, rules, regulations, frameworks, legislation, recommendations, best practice guides and stakeholder pressure can all help shape and govern the behaviours of individuals and practices of organisations. However, the choice between making a correct and incorrect ethical decision does not always mean the choice between abiding by or breaking the law. In examples such as the ones discussed in this book, the correct ethical decision can be arrived at relatively easily. However, the choice between many decisions is often not a legal one, with all choices being lawful, but primarily an ethical one, with some choices more ethically correct and others considered unethical.

The lack of ethical consideration in decision making by some construction professionals can lead to an increase in corruption, money laundering, bribery, unfair competitive practices, discrimination, the de-skilling of professionals, poor health and safety adherence, increased accident frequency, increased dispute prevalence and a decreased trust in the industry for current and potential future professionals. While these disadvantages are not widespread, nor are they experienced on a mass scale, incidents of unethical practice do occur within the construction industry and it is only through the better education of current and future professionals that incidents like this can be reduced and eliminated. The aim of this book was to provide an ethical overview of the construction industry so that all professionals who operate within it, and those planning on entering the industry in the future, are aware of, and can apply, ethical standards to their own behaviour and the wider practices of their organisations.

This book was written in response to two main issues. To serve as a guide as to how and why ethically correct decisions should be made in the variety of situations that professionals face and to illustrate how to practically apply theoretical ethical frameworks. Through the concepts, themes and ideas discussed in this book, a proactive approach to ethical decision making is encouraged to ensure professionals are properly prepared for situations that may arise, requiring ethical consideration and decision making. Hopefully, this book illustrates how ethical considerations can form a proactive framework to inform behaviours and ensure professionals are aware of the correct ethical decisions to make in a variety of situations.

Index

Note: bold page numbers indicate tables; italic page numbers indicate figures.